D1525042

HIGHLY LINEAR INTEGRATED WIDEBAND AMPLIFIERS
Design and Analysis Techniques for Frequencies from Audio to RF

THE KLUWER INTERNATIONAL SERIES IN ENGINEERING AND COMPUTER SCIENCE

ANALOG CIRCUITS AND SIGNAL PROCESSING
Consulting Editor: **Mohammed Ismail**. *Ohio State University*

Related Titles:

DYNAMIC TRANSLINEAR AND LOG-DOMAIN CIRCUITS: *Analysis and Synthesis*, Jan Mulder, Wouter A. Serdijn, Albert C. van der Woerd, Arthur H. M. van Roermund; ISBN: 0-7923-8355-9

DISTORTION ANALYSIS OF ANALOG INTEGRATED CIRCUITS, Piet Wambacq, Willy Sansen; ISBN: 0-7923-8186-6

NEUROMORPHIC SYSTEMS ENGINEERING: *Neural Networks in Silicon*, edited by Tor Sverre Lande; ISBN: 0-7923-8158-0

DESIGN OF MODULATORS FOR OVERSAMPLED CONVERTERS, Feng Wang, Ramesh Harjani, ISBN: 0-7923-8063-0

SYMBOLIC ANALYSIS IN ANALOG INTEGRATED CIRCUIT DESIGN, Henrik Floberg, ISBN: 0-7923-9969-2

SWITCHED-CURRENT DESIGN AND IMPLEMENTATION OF OVERSAMPLING A/D CONVERTERS, Nianxiong Tan, ISBN: 0-7923-9963-3

CMOS WIRELESS TRANSCEIVER DESIGN, Jan Crols, Michiel Steyaert, ISBN: 0-7923-9960-9

DESIGN OF LOW-VOLTAGE, LOW-POWER OPERATIONAL AMPLIFIER CELLS, Ron Hogervorst, Johan H. Huijsing, ISBN: 0-7923-9781-9

VLSI-COMPATIBLE IMPLEMENTATIONS FOR ARTIFICIAL NEURAL NETWORKS, Sied Mehdi Fakhraie, Kenneth Carless Smith, ISBN: 0-7923-9825-4

CHARACTERIZATION METHODS FOR SUBMICRON MOSFETs, edited by Hisham Haddara, ISBN: 0-7923-9695-2

LOW-VOLTAGE LOW-POWER ANALOG INTEGRATED CIRCUITS, edited by Wouter Serdijn, ISBN: 0-7923-9608-1

INTEGRATED VIDEO-FREQUENCY CONTINUOUS-TIME FILTERS: *High-Performance Realizations in BiCMOS*, Scott D. Willingham, Ken Martin, ISBN: 0-7923-9595-6

FEED-FORWARD NEURAL NETWORKS: *Vector Decomposition Analysis, Modelling and Analog Implementation*, Anne-Johan Annema, ISBN: 0-7923-9567-0

FREQUENCY COMPENSATION TECHNIQUES LOW-POWER OPERATIONAL AMPLIFIERS, Ruud Easchauzier, Johan Huijsing, ISBN: 0-7923-9565-4

ANALOG SIGNAL GENERATION FOR BIST OF MIXED-SIGNAL INTEGRATED CIRCUITS, Gordon W. Roberts, Albert K. Lu, ISBN: 0-7923-9564-6

INTEGRATED FIBER-OPTIC RECEIVERS, Aaron Buchwald, Kenneth W. Martin, ISBN: 0-7923-9549-2

MODELING WITH AN ANALOG HARDWARE DESCRIPTION LANGUAGE, H. Alan Mantooth, Mike Fiegenbaum, ISBN: 0-7923-9516-6

LOW-VOLTAGE CMOS OPERATIONAL AMPLIFIERS: *Theory, Design and Implementation*, Satoshi Sakurai, Mohammed Ismail, ISBN: 0-7923-9507-7

ANALYSIS AND SYNTHESIS OF MOS TRANSLINEAR CIRCUITS, Remco J. Wiegerink, ISBN: 0-7923-9390-2

COMPUTER-AIDED DESIGN OF ANALOG CIRCUITS AND SYSTEMS, L. Richard Carley, Ronald S. Gyurcsik, ISBN: 0-7923-9351-1

HIGH-PERFORMANCE CMOS CONTINUOUS-TIME FILTERS, José Silva-Martínez, Michiel Steyaert, Willy Sansen, ISBN: 0-7923-9339-2

SYMBOLIC ANALYSIS OF ANALOG CIRCUITS: *Techniques and Applications*, Lawrence P. Huelsman, Georges G. E. Gielen, ISBN: 0-7923-9324-4

DESIGN OF LOW-VOLTAGE BIPOLAR OPERATIONAL AMPLIFIERS, M. Jeroen Fonderie, Johan H. Huijsing, ISBN: 0-7923-9317-1

STATISTICAL MODELING FOR COMPUTER-AIDED DESIGN OF MOS VLSI CIRCUITS, Christopher Michael, Mohammed Ismail, ISBN: 0-7923-9299-X

HIGHLY LINEAR INTEGRATED WIDEBAND AMPLIFIERS
Design and Analysis Techniques for Frequencies from Audio to RF

by

Henrik Sjöland
Lund University, Sweden

KLUWER ACADEMIC PUBLISHERS
Boston / Dordrecht / London

Distributors for North, Central and South America:
Kluwer Academic Publishers
101 Philip Drive
Assinippi Park
Norwell, Massachusetts 02061 USA
Telephone (781) 871-6600
Fax (781) 871-6528
E-Mail <kluwer@wkap.com>

Distributors for all other countries:
Kluwer Academic Publishers Group
Distribution Centre
Post Office Box 322
3300 AH Dordrecht, THE NETHERLANDS
Telephone 31 78 6392 392
Fax 31 78 6546 474
E-Mail <orderdept@wkap.nl>

 Electronic Services <http://www.wkap.nl>

Library of Congress Cataloging-in-Publication Data

Sjöland, Henrik, 1972-
    Highly linear integrated wideband amplifiers : design and analysis techniques for frequencies from audio to RF / by Henrik Sjöland.
      p. cm. -- (The Kluwer international series in engineering and computer science ; SECS 490)
    Includes bibliographical references and index.
    ISBN 0-7923-8407-5 (alk. paper)
    1. Broadband amplifiers--Design and construction. 2. Linear integrated circuits. 3. Power amplifiers. I. Title. II. Series.
TK7871.58.B74S36   1999   98-48437
621.3815'35--dc21   CIP

Copyright © 1999 by Kluwer Academic Publishers

All rights reserved. No part of this publication may be reproduced, stored in a retrieval system or transmitted in any form or by any means, mechanical, photocopying, recording, or otherwise, without the prior written permission of the publisher, Kluwer Academic Publishers, 101 Philip Drive, Assinippi Park, Norwell, Massachusetts 02061

Printed on acid-free paper.

Printed in the United States of America

Contents

List of Figures	ix
Preface	xiii
Acknowledgements	xv

1 Introduction   **1**
- 1.1 Wideband IF Amplifiers 3
- 1.2 RF Power Amplifiers 5
- 1.3 Audio Power Amplifiers 6
- 1.4 References 8

2 Integrated Transistors and Amplifiers   **9**
- 2.1 Integrated Analog Electronics in Short 9
- 2.2 Amplifiers in General 11
- 2.3 Transistors 14
  - 2.3.1 MOS Transistors 14
    - 2.3.1.1 Small Signal Model 16
    - 2.3.1.2 Dynamic Effects and Noise 17
    - 2.3.1.3 Short Channel Effects 19
  - 2.3.2 Bipolar Transistors 21
- 2.4 References 24

3 Amplifier Linearization Techniques   **25**
- 3.1 Negative Feedback 26
- 3.2 Feed-Forward 27
- 3.3 Predistortion 29
- 3.4 Cancellation 31
- 3.5 References 33

4 Advanced Feedback Techniques 35
- 4.1 The Asymptotic-Gain Model 35
- 4.2 Stability 36
- 4.3 Phase-Compensation Techniques...39
- 4.4 Feedback Boosting..41
  - 4.4.1 The Topology 42
  - 4.4.2 Analysis..44
  - 4.4.3 Detailed Stability Analysis..46
  - 4.4.4 Experiment and Simulations 47
  - 4.4.5 Other Topologies 48
  - 4.4.6 Distortion Cancellation at Signal Frequencies.........49
- 4.5 References 50

5 Output Stages 51
- 5.1 Classes of Operation 52
- 5.2 Low Voltage Output Stages..56
  - 5.2.1 CMOS Output Stages..57
  - 5.2.2 Bipolar Output Stages 60
- 5.3 References 62

6 Analysis and Measurement of Distortion 65
- 6.1 Computer Simulation of Distortion 65
- 6.2 Distortion Measurement..66
- 6.3 Intermodulation Distortion related to THD 67
  - 6.3.1 Models of the Nonlinear Amplifier 67
  - 6.3.2 Input Signals...69
  - 6.3.3 Static Distortion 72
    - 6.3.3.1 Static Clipping 72
    - 6.3.3.2 Static Distortion Before Clipping 73
  - 6.3.4 Dynamic Distortion 76
    - 6.3.4.1 Slew-Rate Clipping.......................................77
    - 6.3.4.2 Dynamic Distortion Before Clipping..........80
  - 6.3.5 The Method 81
  - 6.3.6 Example...81
  - 6.3.7 Numerical Experiments..82
    - 6.3.7.1 Static Nonlinearity Only 82
    - 6.3.7.2 Both Dynamic and Static Nonlinearity.......83
- 6.4 References 85

7 Audio Power Amplifiers   87

- 7.1 Requirements of Audio Power Amplifiers 87
  - 7.1.1 General Requirements 87
  - 7.1.2 Special Requirements of Integrated Battery-Pow 88
- 7.2 Class of Operation..89
  - 7.2.1 Class AB 89
  - 7.2.2 Class S 89
- 7.3 CMOS vs. Bipolar..90
- 7.4 Two CMOS Class AB Designs 91
  - 7.4.1 Output Stages 91
    - 7.4.1.1 Output Stage with Blomley Topology 92
    - 7.4.1.2 A Low Voltage Output Stage 94
  - 7.4.2 Phase-Compensation 97
    - 7.4.2.1 Phase-Compensation of the Blomley 97
    - 7.4.2.2 Phase-Compensation of the Low Voltage ...99
  - 7.4.3 Input Stages 101
    - 7.4.3.1 Blomley Amplifier Input Stage.................101
    - 7.4.3.2 Low Voltage Input Stage 102
  - 7.4.4 Simulation and Measurement Results....................103
    - 7.4.4.1 The Blomley Amplifier Results 103
    - 7.4.4.2 The Low Voltage Amplifier Results 106
  - 7.4.5 Layout and Chip Photos..109
- 7.5 References..111

8 Wideband IF Amplifiers   113

- 8.1 Performance Requirements 113
- 8.2 The Topology 114
- 8.3 Output Stages 115
  - 8.3.1 CMOS Output Stages...116
  - 8.3.2 Bipolar Output Stage...117
- 8.4 Input Stages...118
  - 8.4.1 CMOS Input Stage 118
  - 8.4.2 Bipolar Input Stage 119
- 8.5 Middle Stage 120
- 8.6 Common-Mode Feedback..121
- 8.7 Simulations and Measurements 122
  - 8.7.1 Schematics 123
  - 8.7.2 Results 125
- 8.8 Chip Photos..126
- 8.9 References..128

9 Inductorless RF CMOS Power Amplifiers 129

9.1 Requirements on RF Power Amplifiers 129
9.2 CMOS RF Power Amplifiers using Inductors 130
9.3 An Inductorless CMOS RF Power Amplifier 132
   9.3.1 Amplifier Topology 132
   9.3.2 Output Stage 133
   9.3.3 Driver Stage 135
   9.3.4 Input Stage 137
   9.3.5 Simulations and Measurement Results 137
   9.3.6 Conclusions 141
   9.3.7 Layout and Chip Photo 142
9.4 References 143

10 Layout Aspects 145

10.1 Passive Devices 146
   10.1.1 Resistors 146
   10.1.2 Capacitors 148
   10.1.3 Inductors 150
10.2 Active Devices 152
   10.2.1 MOS Transistors 152
      10.2.1.1 The Layers and Their Physical Corresp 152
      10.2.1.2 Layout for Low Noise 154
      10.2.1.3 Layout for High Speed 154
      10.2.1.4 Layout Techniques for Wide Transistors .. 155
      10.2.1.5 High Current Layout 157
   10.2.2 Bipolar Transistors 157
   10.2.3 Matching of Transistors 158
10.3 References 159

Index 161

List of Figures

1.1	The receiver side of a conventional base station	3
1.2	Base station with wideband IF amplifier	3
1.3	Example of channels with different power	4
1.4	Spectral widening disturbing adjacent channels	5
2.1	The output spectrum when the input is (a) a sinusoid	13
2.2	The intercept diagram showing 3rd order intercept and compress	13
2.3	Enhancement MOS transistor symbols (a) N (b) P	15
2.4	Simple small signal schematic of an MOS transistor	17
2.5	Velocity saturation	19
2.6	Effect of velocity saturation on f_t	20
2.7	Symbols for bipolar transistors (a) npn (b) pnp	21
2.8	A simple small-signal equivalent circuit of a bipolar transistor	23
3.1	The elementary feedback model	26
3.2	Basic feed-forward configuration	27
3.3	Feed-forward configuration with compensation for time-delays	28
3.4	(a) The principle of predistortion (b) An MOS current mirror	29
3.5	Predistortion based on model amplifier	30
3.6	(a) The principle of cancellation (b) A CMOS inverter	31
4.1	(a) Example of Nyquist diagram. (b) Enlarged view around	37
4.2	Example of Bode diagram. (a) Asymptotic (b) With gain and	38
4.3	Illustration of different phase-compensation methods	40
4.4	A three stage amplifier topology with double nested Miller	41
4.5	The basic inverting topology	42
4.6	The complete inverting topology	43
4.7	The boost in the feedback (extra loop gain)	45
4.8	Measured THD with and without feedback boosting	47
4.9	Simulated THD with and without feedback boosting	48
4.10	(a) A non-inverting topology (b) A floating negative resistance	49
5.1	The tail currents of the output stage	52
5.2	The tail currents vs. I_{out} for different classes of operation	53
5.3	Maximum efficiency for class B and push-pull class A vs. amplit	55
5.4	The connection of the output devices in a low voltage output stage	58
5.5	Example of ultimate low voltage push-pull output stage with	59
5.6	Example of low voltage topology with complementary amplifiers	59
5.7	Connection of the output devices in a low voltage bipolar amplifier	60
5.8	Example of low voltage bipolar output stage	60

List of Figures

5.9	Connection of output devices in an npn-only bipolar process	61
5.10	Example of npn-only push-pull stage	61
6.1	Static characteristic	68
6.2	The probability density function of a sinusoid with amplitude A	70
6.3	The probability density function of the Gaussian wideband signal	71
6.4	The spectral density of the wideband signal	71
6.5	THD/IM vs. A/σ and n	75
6.6	THD/IM vs. A and n when $A/\sigma = 2.5$	76
6.7	Slew-rate clipping scenario	78
6.8	Amplifier model for the numerical experiment	83
7.1	A simplified schematic of a class S output stage	89
7.2	The Blomley topology	92
7.3	Schematic of the Blomley output stage	93
7.4	The basic topology of the low voltage output stage	95
7.5	Different methods to establish V_{B2} and V_{B3}	96
7.6	Simplified frequency responses of (a) the output stage	97
7.7	The local feedback network of the output stage	98
7.8	The demanded gain and the open-loop gain of the output stage	99
7.9	The output stage with reversed nested Miller compensation	99
7.10	Voltage gain A and feedback factor β of the output stage	100
7.11	The input stage of the Blomley amplifier	101
7.12	The frequency response of the input stage	102
7.13	The low voltage input stage	102
7.14	Total schematic of the power amplifier with Blomley topology	104
7.15	Total schematic of the low voltage audio power amplifier	106
7.16	THD at 1kHz 8Ω vs output amplitude and supply voltage	107
7.17	THD vs. frequency at $1V_{pp}$ in 8Ω and 1.5V supply	108
7.18	Micrograph of the 8Ω amplifier with Blomley topology	110
7.19	Micrograph of the 1.5V audio power amplifier	110
8.1	Topology of the wideband IF amplifiers	115
8.2	The CMOS output stage	116
8.3	The CMOS output stage including bias circuit	117
8.4	The bipolar output stage	117
8.5	The CMOS input stage	118
8.6	Two different bipolar input stages	119
8.7	CMOS and bipolar middle stage	120
8.8	CMOS common-mode feedback circuit	121
8.9	Bipolar common-mode feedback circuit	122
8.10	The entire CMOS schematic with device parameters	123
8.11	The entire bipolar schematic with parameters	124
8.12	Microphotograph of the CMOS wideband IF amplifier	127
8.13	Microphotograph of the bipolar wideband IF amplifier	127

List of Figures

9.1	Topology of the inductorless CMOS RF power amplifier	132
9.2	Simplified schematic of the output stage	133
9.3	The 3rd order harmonic distortion vs. V_{gseff0} and V_{out}	135
9.4	The driver stage	136
9.5	The total schematic with device parameters	137
9.6	Simulated frequency response	138
9.7	Measured frequency response	139
9.8	Measured intercept diagrams	140
9.9	Measured power added efficiency (PAE)	141
9.10	Microphotograph of the RF power amplifier	142
10.1	Example of resistor layout	147
10.2	Layout of two matched resistors	148
10.3	Simple model of resistor with parasitic capacitance	148
10.4	Layout of two matched capacitors	149
10.5	Simple model of capacitor with parasitics	150
10.6	Layout of an integrated inductor	151
10.7	Simple NMOS transistor layout and the corresponding cross-se	153
10.8	Example of finger layout	155
10.9	Example of waffle-iron layout	156
10.10	High current MOS transistor in a process with two metal layers	158

Preface

This book deals with techniques for designing integrated wideband amplifiers with high linearity. It is intended for professional designers of integrated amplifiers and graduate students in this field. Amplifiers with operating frequencies from Audio to RF are covered.

This book is based on my Ph.D. thesis, which was published in November 1997. The aim of the project, supervised by professor Sven Mattisson, was to design integrated wideband amplifiers with high linearity for different applications.

In the book an enhanced feedback configuration called feedback boosting is described. It is capable of very low distortion. Theoretically a complete distortion cancellation can be achieved. Also described is a statistical method for relating the intermodulation distortion of a wideband signal to the total harmonic distortion (THD) of a single tone. The motivation for this is that the THD, as opposed to the intermodulation distortion, is easy to measure and use as a design parameter. This method can handle both static and dynamic nonlinearities.

High-performance amplifier designs are also presented, complete with measurement results and chip photos. The different designs cover operating frequencies from audio to RF. Two fully integrated CMOS class AB power amplifiers are described. They are capable of directly driving an 8Ω loudspeaker. One of the amplifiers can also operate on a supply voltage as low as 1.5V. Two wideband amplifiers are also presented, one in CMOS and one bipolar. They have high linearity from DC to 20MHz. Finally, a fully integrated CMOS RF power amplifier is described. It is built without using inductors, to investigate what performance then can be achieved.

Lund, 1998
Henrik Sjöland

Acknowledgements

I would like to thank my advisor professor Sven Mattisson, without whom I would not have been able to perform the work on which this book is based. I would also like to thank my parents and all colleagues and friends at the Department of Applied Electronics for their help and encouragement. I am particularly grateful to Dr. Lars Sundström for his support during the last year. I would also like to thank Paula, my fiancée for her patience.

I am also grateful to professor Mohammed Ismail, who suggested that I would write this book based on my Ph.D. thesis, on which he was the opponent.

Financial support has been received from the Swedish National Board for Industrial and Technical Development (NUTEK).

Some of the material has appeared in IEEE Transactions on Circuits and Systems and IEEE Journal of Solid-State Circuits. I am very grateful to IEEE for allowing me to reuse this material, which is of vital importance for this book.

Chapter 1

Introduction

The amplifier is one of the most important electronic building blocks. Input signals to a system coming from sensors such as antennas or microphones are generally weak, while the output signals need to be strong. It is therefore necessary to provide signal gain between input and output by amplifiers.

Most people associate amplifiers with audio power amplifiers driving loudspeakers. This is, however, just one of the numerous amplifier applications. Almost all electronic devices contain amplifiers. For instance a radio, TV and a mobile phone receive a very weak signal from an antenna, which must be amplified. Even when communicating through a wired telephone the signal from the microphone must be amplified. If it were not, the signal would be drowned by noise on the long wires to the telephone exchange.

An ideal amplifier performs a linear signal transfer from the source to the load. If the transfer is not linear, signal components at frequencies not present at the source will be fed to the load. These undesired signals can be related to as nonlinear distortion and are generated both inside and outside the operating frequency band.

When building an amplifier, active devices must be used. Since the present active electronic devices are inherently nonlinear, building an amplifier with a high degree of linearity is difficult. Methods to improve the linearity have therefore been developed. When improving the linearity, it is important not to harm the performance in some other respect that might make the amplifier unusable in the intended application.

There are several methods to achieve a high degree of linearity. Which one to use depends on the operating conditions and requirements of the application. Each time an amplifier is needed, different methods must be compared to find the most suitable one.

When a wideband amplifier with frequency dependent nonlinearity operates on a wideband signal, the output signal gets very complex. The (intermodulation) distortion can be regarded as noise. The signal to noise ratio becomes very difficult to measure and simulate. In this book a novel method is therefore presented that relates the distortion of a wideband signal to that of a single tone. The distortion of a single tone can be described by the THD-figure (Total Harmonic Distortion) and is easy to both simulate and measure.

In this book integrated wideband amplifiers operating at three different frequency regions are investigated: Audio, IF and RF. For each frequency region one application with a need for high linearity has been identified.

The first application is integrated wideband IF amplifiers intended for simultaneous amplification of several channels in base stations for mobile communications. By amplifying several channels in one amplifier, the chip area and power consumption can be reduced. High linearity is required to prevent intermodulation products from strong channels from blocking out weak channels. When designing the IF amplifiers it was necessary to have simple criteria for the required linearity. For this, the method relating intermodulation noise to THD was used.

The second application is integrated RF power amplifiers. It would be very attractive to be able to integrate an RF power amplifier in CMOS, so that an entire mobile phone could be built on one chip. With increased data rates the frequency spectrum must be used more efficiently. To achieve this, modulation schemes with a non-constant envelope have started to appear. When the envelope is not constant, the power amplifiers must have high linearity. Present systems are, for instance, IS-54 (USA) [1] and TETRA (Europe) [2]. In the future, systems based on OFDM and CDMA might become popular. Especially OFDM requires very high linearity.

The third application is fully integrated CMOS audio power amplifiers. By using such an amplifier, a system built on a chip can drive loudspeaker loads directly. High linearity is required to achieve high sound quality.

An enhanced feedback scheme is also presented. It enables a complete nonlinearity cancellation, which is not possible with conventional feedback. A complete cancellation requires, however, perfect matching and can occur at just one frequency.

1.1 Wideband IF Amplifiers

A base station for mobile communications is typically designed to serve several channels simultaneously. The typical setup for the receiver side is to use one IF amplifier preceded by one downconversion mixer and one narrowband filter per channel, figure 1.1.

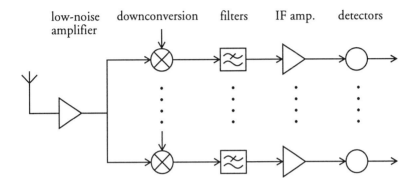

Figure 1.1: The receiver side of a conventional base station

An alternative is to use just one mixer, filter and IF amplifier for several or all channels, figure 1.2.

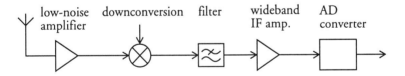

Figure 1.2: Base station with wideband IF amplifier

A wideband IF amplifier and filter must then be used. The advantage is the possibility to reduce the chip area and power consumption.

The power may vary a lot between different channels. It may be strong in some and others at the sensitivity limit (noise floor) in others, see figure 1.3.

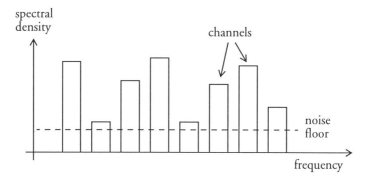

Figure 1.3: Example of channels with different power

When a nonlinear amplifier is fed with a wideband signal, intermodulation noise will occur [3]. The total noise is the intermodulation noise added to other essentially signal independent noise sources. At high signal levels the intermodulation noise is likely to become dominant. The intermodulation noise due to the strong channels must be sufficiently low, so that the weak channels are not blocked out. This requires a high degree of linearity.

It is not apparent how to specify the linearity requirements. A specification that is easy to verify in simulations and measurements is preferable. The total harmonic distortion (THD) is easy to measure and simulate since a single sinusoid is used as the input signal. In this book a method is presented that relates the intermodulation noise of a wideband signal to THD. To accomplish this a statistical approach is employed. Both static and dynamic (frequency dependent) nonlinearities are handled.

The static case has been treated in [3] using a wideband signal consisting of a large number of sinusoids. Amplifier clipping is not considered, however, resulting in very pessimistic results for high order distortion. There is no amplitude limit for a wideband signal, but the probability of extreme amplitudes is low. If these large amplitudes are fed to a high order polynomial, associated with high order distortion, the resulting distortion will be enormous. In reality the distortion is limited in those cases by amplifier clipping. For low order nonlinearities the results correspond to those of our method.

In chapter 8 it is investigated how a highly linear wideband amplifier can be designed both in CMOS and bipolar technology. A topology suitable for both technologies is found. It would also be possible to build an amplifier in BiCMOS using that topology, and thereby combine the advantages of CMOS and bipolar.

1.2 RF Power Amplifiers

When trying to integrate an entire mobile telephone on a CMOS chip, the RF power amplifier is one of the major challenges. The amplifier must be power efficient and linear at very high operating frequencies. The power efficiency is important since mobile phones are battery operated.

High linearity is required with many modern modulation techniques. If the power amplifier is nonlinear, the signal might become distorted to such an extent that it can not be reconstructed and detected at the receiver. This problem is caused by intermodulation distortion inside the channel bandwidth.

When a channel passes a nonlinearity, its spectrum is widened due to intermodulation products outside the channel bandwidth [4]. Nonlinearity can thus also cause adjacent channel interference, see figure 1.4.

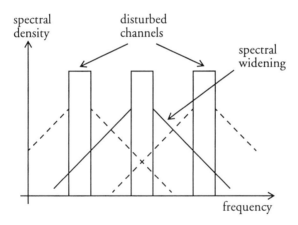

Figure 1.4: Spectral widening disturbing adjacent channels

In order to reach operating frequencies of 2GHz using a standard CMOS process, coils must be employed to circumvent the low speed of the transistors. Parasitic capacitances can then, together with coils, form parallel resonance circuits. The impedance magnitude of such a resonance circuit can be larger than that of a capacitor alone, resulting in an increased gain. The maximum increase of the impedance occurs at the resonance frequency, where it is equal to the Q-factor of the coil. The problem is that coils with high Q-factor can not be built in standard CMOS processes.

With future CMOS processes the transistors will probably be fast enough to operate on 2GHz without help from coils. In chapter 9 it is investigated, therefore, how fast and linear an amplifier can be made without coils using a standard 0.8μm CMOS process. An amplifier capable of 300MHz operation was built. Based on that, the required channel length for 2GHz operation is then estimated to be about 0.2μm.

1.3 Audio Power Amplifiers

When building a system on a CMOS chip, input and output signals of various types might be needed. An example is an output capable of driving a loudspeaker. In chapter 7 low voltage CMOS audio power amplifiers capable of driving 8Ω loudspeakers are built using different topologies.

Audio power amplifiers must be able to deliver relatively large currents. High power-efficiency is therefore important. It is hard to achieve this in combination with high linearity. Furthermore, if the design is to be fully integrated, large variations of load impedance must be handled. In chapter 7 the amplifiers are designed to manage up to 100nF load capacitance without stability problems. The total capacitance on the chip must be less than about 10pF, if a standard CMOS process is used and the area is not to become excessive. The quotient of the load to the compensation capacitance is therefore larger than 10,000, which makes it difficult to phase-compensate such an amplifier without spoiling the linearity.

Due to the limited supply voltage, the output stage must be able to deliver as large an output voltage swing as possible. Even if the output stage is well designed, the swing can not exceed the supply voltage. The theoretical maximum power in 8Ω at a 5V supply thus becomes 390mW(rms). In a bridge connection this increases to 1.56W(rms). It will thus not be possible to play loud music or to reproduce sound effects on films at realistic levels, but for some applications, such as background music or earphone listening, the power is sufficient.

High power-efficiency is important in order to avoid the need for additional cooling and to get a long battery life in portable applications. To achieve this the output stage must operate in a power efficient class, such as B or AB, and the current consumption in the rest of the amplifier stages must be kept low. Some theory on class AB stages and how to phase-compensate them is found in [5].

It is desirable if one supply voltage can be used for all the subsystems of a system on a chip. It is therefore an advantage if the power amplifier is flexible in the supply voltage requirement. In chapter 7 it is examined how low the supply voltage of a class AB CMOS audio power amplifier can be, and an amplifier with a sup-

1.3 Audio Power Amplifiers

ply voltage range from 1.5V to 5V is presented. The maximum output power is, of course, reduced at low supply voltages.

Negative feedback is effective to linearize audio power amplifiers. The frequency is low, so the gain of the transistors is large, which makes it possible to realize large loop gains. The only problem is the variation of the load impedance, making the phase-compensation difficult. The linearity will be highest if the amplifier is made as linear as possible even before feedback is applied. In the designs of chapter 7 the output devices are part of current mirrors with good current transfer linearity.

1.4 References

[1] Electronic Industries Association. *Dual-mode subscriber equipment - network equipment compatibility specification.* Interim Standard 54, December 1989

[2] J. F. Wilson, 'The TETRA system and its requirements for linear amplification', *IEE Colloquium on 'Linear RF Amplifiers and Transmitters', (Digest No:1994/089)*, pp. 4/1-7, April 1994

[3] R. A. Brockbank, C. A. A. Wass, 'Non-Linear Distortion in Transmission Systems', *J. Inst. Elec. Engrs.,* part III, pp. 45-56, March 1945

[4] J. Boccuzzi, 'Performance Evaluation of Non-Linear Transmit Power Amplifiers for North American Digital Cellular Portables', *IEEE Trans. on Vehicular Technology,* vol. 44, no. 2, pp. 220-228, May 1995

[5] J. H. Huijsing, R. J. van der Plassche and W. Sansen (editors), *Analog Circuit Design*, pp.113-138 (by R. Castello), Kluwer Academic Publishers, 1993

Chapter 2

Integrated Transistors and Amplifiers

This chapter is an introduction to integrated amplifiers. The first section discusses advantages and drawbacks of integrated analog electronics. In section 2.2 the ideal amplifier is defined, followed by a description of parameters that constitute the deviation from the ideal amplifier. The use of both bipolar and CMOS devices is considered in this book, and section 2.3 gives an introduction to their behaviour and how to model them mathematically.

2.1 Integrated Analog Electronics in Short

Integrated circuits have several advantages over discrete circuits. Perhaps the most important advantage is cost, since integrated circuits can be mass-manufactured at low cost per chip. In addition, integrated circuits are smaller and often consume less power than discrete, which is important in portable equipment.

In an integrated circuit each transistor can often be custom designed for its task. Moreover, there are no economical objections against complicated circuits with a lot of transistors, as in integrated circuits the area of the chip and not the number of transistors determines the cost.

Passive devices are less suitable for integration for several reasons. The size of a capacitor is limited to a few pF, if the area is not to be excessive. For the same reason, the resistance and current handling capability of a resistor is limited. Inductors up to a few nH can be made, but very low quality factors are achieved in standard processes, typically less than 5.

The absolute accuracy of both the active and passive devices is usually low, but as the devices are manufactured simultaneously on the same chip, the relative accuracy (matching) is usually very high. Circuit topologies requiring good matching are therefore suitable for integration [1].

Better performance of passive components can be obtained if dedicated analog processes are used, but the cost will then increase. The process is to be selected so that the chips are as cheap as possible to manufacture, but still have the required performance. The cheapest family of processes is the standard digital CMOS. Due to development of the processes, the CMOS transistors become smaller and faster every year. CMOS circuits operating at RF (Radio-Frequency) are already being built. These processes are also very suitable for digital circuits, making it possible to integrate analog and digital electronics on the same chip.

Bipolar transistors have, however, some advantages over CMOS. They have higher transconductance for the same current and area consumption. The offset voltage is lower when used as a differential pair in an amplifier input-stage. Moreover, they can be used in translinear circuits [2]. In CMOS it is possible to build 'CMOS translinear circuits'. These, however, have some drawbacks compared to the translinear circuits implemented using bipolar transistors [3]. Bipolar transistors are also slightly faster than the same generation of CMOS. Therefore there are bipolar processes that are used mainly for analog circuits. A disadvantage of many bipolar processes is that only npn-transistors show high performance.

The advantages of CMOS and bipolar processes are combined in BiCMOS processes, in which both bipolar and CMOS devices can be built on the same chip [4]. The drawback, however, is the increased cost.

Before a chip is used, it is usually put in some form of package that protects the chip and enables mounting on a printed circuit board (PCB). At high frequencies, package parasitic impedances result in difficulties getting signals on and off the chip, since the package resembles, more or less, a low-pass filter. Inductances in series with the ground and power supplies can introduce undesired feedback and thereby cause stability problems.

Design of integrated circuits requires the chips to be carefully designed and simulated before they are manufactured, since in principle it is impossible to change anything inside an integrated circuit without remanufacturing.

2.2 Amplifiers in General

An amplifier is a device that transfers a signal from a source to a load. The signal at the source is first divided between the source and the input impedances, then it is multiplied by the gain, and finally it is divided between the output and load impedances. An ideal amplifier is fully characterized by the gain and the input and output impedances.

The desired input and output impedances are often zero (short circuit), infinite (open circuit) or equal to some characteristic impedance. If the signal is a voltage, the input impedance is ideally infinite and the output impedance zero. In the case of a current signal, the opposite is true. In high frequency applications where the input/output is connected to a transmission line, the input/output impedance must often be equal to the characteristic impedance of the transmission line in order to avoid reflections.

As no ideal amplifier exists in reality, additional parameters must be used to describe a real amplifier. Some non-ideal characteristics of a typical amplifier are:

- Generation of noise
- Limited bandwidth
- Possibility of instability
- Non-zero power consumption
- Non-zero physical size (chip area)
- Nonlinearity

I will now give a short description of these non-ideal properties and the most common parameters being used to quantify them.

All amplifiers add noise to the signal. The signal to noise ratio (SNR) is thus smaller at the output of an amplifier than at the input. The noise can have a frequency dependent spectral density. The noise figure is defined as the ratio of input SNR to output SNR [5]. For a noiseless amplifier the noise figure is equal to one, and for all others, larger than one. The noise figure together with the associated source impedance can be used to describe the noise performance. Another way to describe the noise is to transfer all noise sources to equivalents at the input of the amplifier. If both a voltage and a current equivalent noise source are used, the source impedance must not be specified.

The high and low frequency band edges are specified as the frequencies where the small signal gain is reduced by 3dB with respect to its intended value [6].

The amplifier might be unstable for some load (and source) impedances. In that case it must be specified which impedances are allowed.

Active devices, and thereby amplifiers, consume power. The power consumption is an important figure of merit for an amplifier. For power amplifiers the efficiency is often specified. It is also important to know the supply voltage.

For an integrated amplifier the required chip area is often important.

The limits of the output voltage or current are often important parameters. Also the maximum positive and negative time derivatives of the output, the slew-rate, can be important. If the maximum positive and negative values of the slew-rate or output level are different, the smallest one or both can be given. If the demanded output or its time derivative is outside the limits, a gross distortion called clipping occurs. Slew-rate clipping has sometimes been referred to as transient intermodulation distortion (TIM) [7].

As mentioned in chapter 1, nonlinearity is an important, undesired property of amplifiers. To fully characterize the nonlinearity of a dynamic amplifier is extremely difficult and involves Volterra analysis [8]. The simplest distortion measurement is the THD-measurement. A sinewave is fed to the input, and the output harmonics at multiples of the input frequency are measured, figure 2.1a. The THD can then be calculated as [9]:

$$THD = \frac{\sqrt{A_{2f}^2 + A_{3f}^2 + \ldots + A_{nf}^2}}{A_f} \qquad (2.1)$$

where A_{if} is the amplitude of the harmonic at frequency $i \cdot f$. To characterize the amplifier, the THD can be measured at different amplitudes and frequencies. Another measurement is the two-tone measurement, in which two sinewaves at different frequencies are fed to the input simultaneously. The output spectra is then studied, figure 2.1b.

2.2 Amplifiers in General

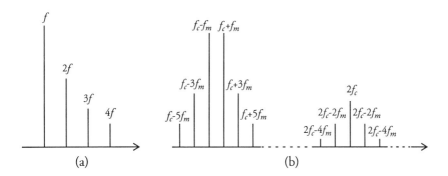

Figure 2.1: The output spectrum when the input is (a) a sinusoid at frequency f, (b) two sinusoids at f_c-f_m and f_c+f_m, respectively

The harmonics at f_c-$3f_m$ and f_c+$3f_m$ are caused by third order nonlinearity.

As a measure of linearity, radio engineers often use the intercept points and 1dB compression point [10]. These are most easily defined by drawing the intercept diagram, where the levels of the output fundamental and harmonics are drawn versus the input level, figure 2.2.

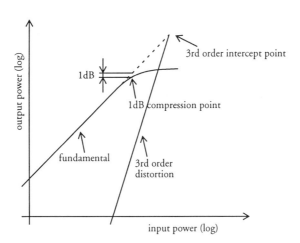

Figure 2.2: The intercept diagram showing 3rd order intercept and compression points

It can also be important to know how close the input and output impedance are to the desired ones. The accuracy of the gain can also be important.

Which properties are important is determined by the specific application in which the amplifier is to be used. Other properties not mentioned here might be important in some situations: for instance, immunity to disturbances on the power supply (PSRR) or fast step response (settling).

2.3 Transistors

In order to provide signal gain, an amplifier must contain at least one active device. In this book MOS and bipolar transistors are used as active devices. When designing an amplifier, device models of various complexity can be used. In this section, simple models suitable for hand calculations are presented. More advanced models are restricted to computer simulations. This section is in no way meant to be complete, as semiconductor physics and device modelling are large fields of research in which several books have been published. Further reading can be found in the references, see for example [1,11,12,13].

2.3.1 MOS Transistors

The MOS transistor is usually modelled with four terminals: gate, bulk, drain and source. The transistor can be made symmetric, so that there is no physical difference between the drain and the source. For both symmetric and non-symmetric transistors, it depends on the driving conditions which terminal is called drain and which is called source.

The current path is between drain and source. If the drain to source voltage is sufficiently large, the current is controlled mainly by the gate to source voltage. The drain voltage has little influence, and the drain is thus a high impedance terminal. The source to bulk voltage has some influence. The bulk is the silicon surrounding the transistor.

There are two types (polarities) of MOS transistors, N and P. The N device conducts when the gate-source voltage is more positive than the threshold voltage V_{Tn}, which is dependent on parameters such as doping concentrations. The source has lower potential than the drain. In an N device electrons are charge carriers, and the current direction is from drain to source. The N device can be used as a current sink. The P device conducts when the gate-source voltage is more negative than the threshold voltage V_{Tp}. The source has higher potential than the drain. In a P device holes are charge carriers, and the current goes from source to drain. The P device can be used as a current source.

2.3 Transistors

If the devices are of enhancement type, the most common device in integrated circuits, they are cut off when the gate-source voltage is zero, that is V_{Tn} is positive and V_{Tp} is negative. One way of drawing the symbols of enhancement N and P devices is shown in figure 2.3:

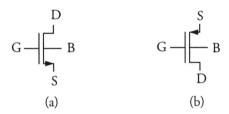

Figure 2.3: Enhancement MOS transistor symbols (a) N (b) P

No direct current, apart from very small leakage currents, passes through the gate, since it is located on an insulating oxide layer. This is one reason for the popularity of MOS, since digital circuits without static power consumption can be designed.

To simplify the following discussion we assume the transistor to be of N type. It is convenient to describe the MOS transistor as operating in different regions, although there are no sharp borders separating them. If the gate-source voltage, V_{gs}, is larger than V_{Tn}, then the transistor is in strong inversion. The strong inversion region can be divided into two subregions, the saturated region and the linear region. If the drain-source voltage, V_{ds}, is larger than $V_{gs}-V_{Tn}$, the transistor is in the saturated region, otherwise, in the linear. In the saturation region the dependence of the current on V_{ds} is small; that is, the drain is a high resistance terminal. Ideally the transistor operates as a voltage controlled current source, where the current is proportional to the square of $V_{gs}-V_{Tn}$. In the linear region the current depends on V_{ds} and is proportional to V_{ds} for small values. The transistor behaves as a voltage controlled resistance. When V_{gs} is smaller than V_{Tn}, the transistor is in weak inversion, and the current is then very small and exponentially dependent on V_{gs}. Finally, when V_{gs} is sufficiently small, the device is off.

Transistors in weak inversion are used in very low power applications. Only strong inversion operation is used in the amplifiers described in this book, where the current, I_d, can be approximated by:

$$\begin{cases} I_d = \frac{1}{2}\mu_{eff} C_{ox} \frac{W}{L}(V_{gs} - V_{Tn})^2(1 + \lambda V_{ds}) & V_{ds} > V_{gs} - V_{Tn} \\ I_d = \mu_{eff} C_{ox} \frac{W}{L}\left((V_{gs} - V_{Tn})V_{ds} - \frac{V_{ds}^2}{2}\right) & V_{ds} < V_{gs} - V_{Tn} \\ V_{Tn} = V_{TO} + \gamma(\sqrt{\phi_B - V_{bs}} - \sqrt{\phi_B}) \\ \mu_{eff} = \frac{\mu_0}{1 + \theta(V_{gs} - V_{Tn})} \end{cases} \quad (2.2)$$

These equations are derived for transistors with long channels. When the dimensions decrease, the equations get more complicated. The minimum channel length of the transistors in this book is 0.8µm, and (2.2) can be used as an approximation.

The current is proportional to W/L, where W is the width and L is the length of the transistor. In an integrated circuit, transistors can be tailored using different widths and lengths.

The mobility μ is larger for electrons than holes ($\mu_n > \mu_p$). In a typical MOS process the quotient is about three, making it necessary for a P device to have about three times the W/L ratio of an N device, if their current drive capability is to be equal.

2.3.1.1 Small Signal Model

The derivatives of the current with respect to the terminal voltages are interesting, as they can be used to calculate the small signal gain and impedances of a circuit. The derivatives are calculated at the operating point of the device when used in the circuit. The variation in the mobility is ignored to get simpler expressions.

2.3 Transistors

$$\begin{cases} g_m = \dfrac{\partial I_d}{\partial V_{gs}} = \mu_n C_{ox} \dfrac{W}{L}(V_{gs} - V_{Tn})(1 + \lambda V_{ds}) = \dfrac{2I_d}{V_{gs} - V_{Tn}} \\ g_d = \dfrac{\partial I_d}{\partial V_{ds}} = \dfrac{1}{2}\mu_n C_{ox} \dfrac{W}{L}(V_{gs} - V_{Tn})^2 \lambda = \dfrac{\lambda I_d}{1 + \lambda V_{ds}} \\ g_{mb} = \dfrac{\partial I_d}{\partial V_{bs}} = \dfrac{g_m \gamma}{2\sqrt{\phi_B - V_{bs}}} \end{cases} \quad (2.3)$$

when $V_{ds} > V_{gs} - V_{Tn}$

$$g_d = \dfrac{\partial I_d}{\partial V_{ds}} = \mu_n C_{ox} \dfrac{W}{L}(V_{gs} - V_{Tn} - V_{ds}) \qquad V_{ds} < V_{gs} - V_{Tn} \qquad (2.4)$$

where g_d is the drain conductance, while g_m and g_{mb} are transconductances. The quotient g_m/g_d is the intrinsic voltage gain of the transistor. It can be noted that the voltage gain seems to become infinite when V_{gs} is equal to V_{Tn}. This does not occur in reality, as the transistor then reaches weak inversion.

The small signal schematic of the transistor can be drawn using the derivatives just found. This is just a linearization of the equations used to model the transistor.

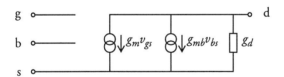

Figure 2.4: Simple small signal schematic of an MOS transistor

2.3.1.2 Dynamic Effects and Noise

So far only static voltages and currents have been considered. If the transistor is to be used with signals, dynamic effects also have to be taken into account. In an MOS transistor these can be modelled using capacitances between all terminals. The capacitances connected to the gate are of particular importance as they determine the transit frequency f_t, which is the frequency where the current gain is equal to unity when the drain is held at a constant potential.

$$f_t = \frac{g_m}{2\pi(c_{gs}+c_{gd}+c_{gb})} = \frac{\mu C_{ox}\frac{W}{L}(V_{gs}-V_{Tn})}{2\pi(c_{gs}+c_{gd}+c_{gb})} \approx \frac{\mu C_{ox}\frac{W}{L}(V_{gs}-V_{Tn})}{2\pi C_{ox}WL} = \frac{\mu(V_{gs}-V_{Tn})}{2\pi L^2} \quad (2.5)$$

f_t is an important figure of merit in determining how fast an amplifier can be using this transistor. As can be seen, f_t depends on the operating point. The transistor length is important for speed. When the gate length is short, f_t is determined by short channel effects, however, and is proportional to $1/L$ instead of $1/L^2$.

If the dependency of mobility on V_{gs} is considered, f_t increases towards a maximum, f_{tmax}, as V_{gs} is increased:

$$f_{tmax} = \frac{\mu_0}{4\pi\theta L^2} \quad (2.6)$$

The resistance of the gate, r_g, causes noise and deteriorates the high frequency performance. The gate is often made of doped polycrystalline silicon, which has a much higher resistivity than metal. To get a low gate resistance, it is necessary to have densely placed connections (contacts) from the gate to metal.

The noise of the MOS transistor is, apart from the thermal noise of the gate resistance, thermal noise from the channel and $1/f$ noise. The thermal noise of the channel referred to the input is:

$$\overline{v_{nT}^2} = \frac{8kT}{3g_m}\Delta f \quad (2.7)$$

At a certain frequency, f, and generator impedance, R_g, the optimum transconductance for low noise, g_{mopt}, is found as:

$$g_{mopt} = \frac{1}{R_g+r_g}\left(\frac{f_t}{f}\right) \quad (2.8)$$

2.3 Transistors

In applications with operating frequencies far below f_t, g_{mopt} often results in an unsuitably large chip area and current consumption. The transconductance must then be selected below the optimum, and a compromise must be made between noise performance, power consumption and chip area.

The noise in transistors with short gate length can be higher due to hot electrons, generated by large electric fields. The hot electron noise has a constant spectral density:

$$\overline{v_{nT}^2} = a\frac{8kT}{3g_m}\Delta f \qquad (2.9)$$

where a is equal to unity in the absence of hot electrons. In the presence of hot electrons a can exceed 7, depending on bias, geometry and temperature [14].

The $1/f$ noise is a problem in low frequency applications, as the spectral density is proportional to $1/f$. This noise decreases with increasing gate area (WL).

2.3.1.3 Short Channel Effects

As the fabrication processes become more sophisticated, CMOS transistors with shorter channel lengths can be realized. With shorter channels the horizontal electric fields increase. Equation (2.2) is only valid if the average speed of the charge carriers is proportional to the electric field. The proportionality constant is called the mobility, μ. At high electric fields, however, the velocity reaches a maximum v_{max}, see figure 2.5.

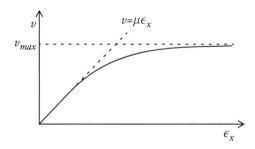

Figure 2.5: Velocity saturation

The drain current becomes:

$$I_d \approx WC_{ox}(V_{gs} - V_{Tn})v_{max} \tag{2.10}$$

The current is a linear instead of a square-law function of V_{gs} and is now independent of the channel length. The maximum velocity is roughly equal for electrons and holes, resulting in nearly the same performance for N and P devices when operating in velocity saturation [11]. The transconductance becomes:

$$g_m = WC_{ox}v_{max} = \frac{I_d}{(V_{gs} - V_{Tn})} \tag{2.11}$$

If (2.11) is compared to (2.3) the only difference is a factor of two. This occurs when the square-law relation (2.2) is differentiated. The transit frequency can now be calculated.

$$f_t = \frac{g_m}{2\pi(c_{gs} + c_{gd} + c_{gb})} \approx \frac{WC_{ox}v_{max}}{2\pi WLC_{ox}} = \frac{v_{max}}{2\pi L} \tag{2.12}$$

When the channel is short, the transit frequency is thus proportional to the inverse of the channel length instead of the channel length squared, figure 2.6.

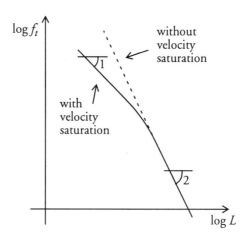

Figure 2.6: Effect of velocity saturation on f_t

2.3 Transistors

Short channel effects make it difficult to predict the speed of future CMOS processes. Further reading about the physics of the MOS transistor can be found in [11,12]. More design oriented is the approach in [1].

2.3.2 Bipolar Transistors

The bipolar transistor is usually modelled using three terminals, as the substrate has little effect. The terminals are base, collector and emitter. The basic principle of a bipolar transistor is that the base is used to control the current between the collector and emitter.

There are two types of bipolar transistors, npn and pnp, corresponding to N and P in MOS. An npn can be used as a controlled current sink and a pnp as a controlled current source. As a pnp is just the complement to an npn, the discussion in the following will be restricted to npn only. In addition, the only bipolar design described in this book uses npn transistors exclusively. The symbols of the bipolar transistors are shown in figure 2.7.

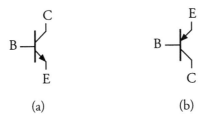

Figure 2.7: Symbols for bipolar transistors (a) npn (b) pnp

In contrast to an MOS transistor, a direct current flows through the control terminal (base). The collector current is related to the base current through the current gain β, which is an important parameter. The base emitter voltage, V_{be}, is related to the base current by an exponential diode current relationship. Depending on the source impedance, the transistor can be controlled either by the base current or the base-emitter voltage. Current control usually results in better linearity, since I_c is more linear as a function of I_b than of V_{be} (exponential).

As long as the collector-emitter voltage is larger than the saturation voltage (about 0.2V), the dependency of the current on the collector voltage is small; that is, the collector is a high impedance node. The collector corresponds to the drain of an MOS device. In the same way, the emitter corresponds to the source and the base to the gate.

If the currents are moderate they follow approximately the relationship given by

$$\begin{cases} I_c = I_S \exp\left(\dfrac{V_{be}}{V_T}\right) \cdot \left(1 + \dfrac{V_{ce}}{V_A}\right) \\ I_b \approx \dfrac{I_S}{\beta_{FM}} \exp\left(\dfrac{V_{be}}{V_T}\right) \end{cases} \qquad (2.13)$$

where V_T is the thermal voltage (which is 26mV at room temperature) and β_{FM} is the maximum value of β. V_A depends on the fabrication process; I_S also on the geometry of the transistor. These simple formulas do not apply for low and high currents.

The small signal parameters are:

$$\begin{cases} g_m = \dfrac{\partial I_c}{\partial V_{be}} = \dfrac{I_c}{V_T} \\ g_\pi = \dfrac{1}{r_\pi} = \dfrac{\partial I_b}{\partial V_{be}} = \dfrac{I_b}{V_T} \approx \dfrac{I_c}{\beta_{FM} V_T} \\ g_o = \dfrac{\partial I_c}{\partial V_{ce}} \approx \dfrac{I_c}{V_A} \end{cases} \qquad (2.14)$$

where g_m is the transconductance, g_π the input conductance, and g_o the output conductance.

The transistor is affected by several parasitic impedances. There are resistances in series with the terminals. The resistance in series with the base, r_b, is particularly harmful. It affects the noise and high frequency behaviour. As in an MOS device, there are also capacitances between the terminals. The transit frequency f_t can be calculated as:

$$f_t = \dfrac{g_m}{2\pi(c_\pi + c_\mu)} \qquad (2.15)$$

where c_π is the base-emitter and c_μ the base-collector capacitance. A difference from MOS devices is that these capacitances are dependent on the terminal voltages (currents) of the transistor. A change of I_c has an effect on the capacitances, as well as on g_m, in such a way that f_t has a maximum at a certain current.

2.3 Transistors

A simple small-signal equivalent circuit can now be drawn:

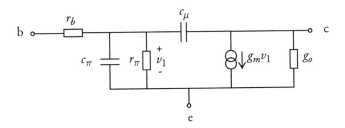

Figure 2.8: A simple small-signal equivalent circuit of a bipolar transistor

There are several noise sources in the bipolar transistor. As for an MOS transistor, a noise optimum exists [15].

$$g_{mopt} = \frac{\sqrt{\beta}}{R_g + r_b} \qquad (2.16)$$

For low-impedance sources bipolar transistors have better noise performance than a corresponding MOS, including $1/f$ noise, if the chip area and power consumption must be low.

Further reading about the physics of the bipolar transistor can be found in [12], whereas more design oriented issues are found in [13].

2.4 References

[1] R. Gregorian and G. C. Themes, *Analog MOS Integrated Circuits for Signal Processing*. Wiley, 1986

[2] C. Toumazou, F. J. Lidgey and D. G. Haigh (editors), *Analog IC Design: the Current-Mode Approach*. Peter Peregrinus Ltd. on behalf of IEE, 1990

[3] W. M. C. Sansen, J. H. Huijsing and R. J. van de Plassche (editors), *Analog Circuit Design*. Kluwer Academic Publishers, 1996

[4] M. Ismail and T. Fiez, *Analog VLSI, Signal and Information Processing*, McGraw-Hill, 1994

[5] J. Davidse, *Analog Electronic Circuit Design*. Prentice Hall, 1991

[6] E. H. Nordholt, *Design of High-Performance Negative-Feedbank Amplifiers*. Delftse Uitgevers Maatschappij, 1993

[7] E. M. Cherry, 'Transient Intermodulation Distortion - Part I: Hard Non-linearity', *IEEE Trans. on Acoustics, Speech, and Signal Processing*, vol. ASSP-29, no. 2, pp. 751-756, Apr 1981

[8] S. Narayanan, 'Application of Volterra Series to Intermodulation Distortion Analysis of Transistor Feedback Amplifiers', *IEEE Trans. on Circuit Theory*, vol. CT-17, no. 4, pp. 518-527, Nov. 1970

[9] E. M. Cherry and D. E. Hooper, *Amplifying Devices and Low-Pass Amplifier Design*. Wiley, 1968

[10] P. H. Young, *Electronic Communication Techniques, third edition*. Macmillan, 1994

[11] Y. P. Tsividis, *Operation and Modeling of the MOS Transistor*. McGraw-Hill, 1987

[12] S. M. Sze, *Physics of Semiconductor Devices, 2nd edition*. Wiley, 1981

[13] P. R. Gray and R. G. Meyer, *Analysis and Design of Analog Integrated Circuits, third edition*. Wiley, 1993

[14] M. J. Buckingham, *Noise in Electronic Devices and Systems*, Wiley, 1983

[15] Y. Netzer, 'The Design of Low-Noise Amplifiers', *Proc. of the IEEE*, vol. 69, no. 6, pp. 728-741, June 1981

Chapter 3

Amplifier Linearization Techniques

If the quiescent current of an amplifier stage is much larger than the maximum signal current, the current dependent small-signal parameters of the transistors will be nearly constant and independent of the signal current. Low distortion can thereby be achieved by using large quiescent currents. This will, however, in most situations result in unacceptably high power consumption. There are several methods, which can be combined, to achieve low distortion without excessive power consumption. Different linearization methods are presented in the subsections of this chapter.

The methods described are negative feedback, feed-forward, predistortion and cancellation. The descriptions are brief, concentrating on principles rather than details. The negative feedback, however, deserves more treatment, as it is very widely used and often requires advanced measures to avoid self-oscillations. Chapter 4 is therefore entirely devoted to more advanced topics regarding negative feedback.

3.1 Negative Feedback

Negative feedback was invented by H. S. Black in the 1920's [1,2]. It is based on a scheme where the error is found by subtracting the output signal, divided by the desired gain, from the input signal of the circuit. This error is fed to the input of the amplifier to be linearized, in such a way that the error at the output is counteracted, see figure 3.1.

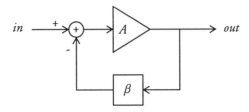

Figure 3.1: The elementary feedback model

The gain from input to output in figure 3.1 is:

$$A_{cl} = \frac{1}{\beta} \cdot \frac{\beta A}{1 + \beta A} \tag{3.1}$$

When βA, the loop gain, is large, A_{cl} is determined by just β, which is usually passive and thus more linear than A. If the gain is independent of A, the amplifier distortion due to nonlinearities in A is eliminated. The larger the loop gain, the more independent is the gain on A, and the more linearized is the amplifier. As β is fixed by the desired gain, we must make A as large as possible to maximize the loop gain. Negative feedback can be regarded as an exchange, where gain is paid for linearity.

Feedback can also be used to increase the accuracy of an amplifier, as β generally is more accurate than A. This is important, as the transistor parameters can have a large spread between different fabrications and a large variation with temperature. For integrated amplifiers particularly high accuracy can be achieved when the gain is dimension-less, as in a voltage or current amplifier. The reason is that the β network then can rely on a quotient between two passive components of the same type, which is the most accurate that can be built on an integrated circuit.

A problem with feedback is that the amplifier to be linearized is fed by an error signal, and as the amplifier needs an input signal, the error can not be completely eliminated. A solution to this problem is to use feedback boosting, which is described in section 4.4. The most important problem with feedback is the risk of self-oscillations. They can occur since the output is connected to the input, thereby forming a loop. To avoid them it is necessary to have control of the phase-shifts in the loop, that is, of A and β. This is accomplished by phase-compensation, which is treated in chapter 4. Another problem is that at high frequencies (RF) it is difficult to get enough loop gain to linearize an amplifier sufficiently.

Negative feedback is treated in several books on electronics and control theory. An early and very important book is [3].

3.2 Feed-Forward

Feed-forward was, like feedback, invented by H. S. Black in the 1920's [1,4]. The invention of feed-forward preceded that of feedback by several years. In the feed-forward scheme the error is subtracted directly from the amplifier output, instead of being fed to the amplifier input. This enables a complete cancellation of the nonlinearity, and there are no stability problems [5]. The basic feed-forward configuration is shown in figure 3.2.

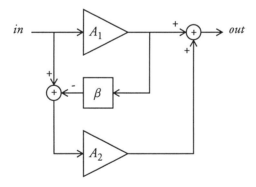

Figure 3.2: Basic feed-forward configuration

The amplifier with gain A_1 is the main amplifier to be linearized and the one with gain A_2 is an auxiliary amplifier that amplifies the error signal. The gain from the input to the output is:

$$A_{tot} = A_1 + A_2 - \beta A_1 A_2 \qquad (3.2)$$

If A_2 is selected according to (3.3), the error of the main amplifier is cancelled.

$$A_2 \beta = 1 \qquad (3.3)$$

When (3.3) is satisfied, only the auxiliary amplifier will contribute nonlinear distortion. If this is a class A amplifier carrying small signals, the distortion levels can be very small. If A_1 is selected equal to the total gain ($1/\beta$), the input signal of the auxiliary amplifier is minimized.

Especially in high frequency applications, it might be necessary to insert time delay blocks to account for the time delays of the amplifiers, see figure 3.3. The purpose of delay τ_1 is to compensate for the delay of the main amplifier. In the same way τ_2 compensates for the delay of the auxiliary amplifier.

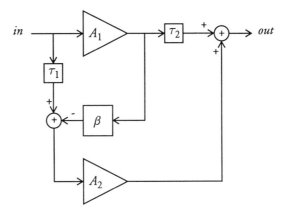

Figure 3.3: Feed-forward configuration with compensation for time-delays

One major problem, especially in integrated circuits, is how to realize the summation of signals at the output. One difficulty is that leakage from the output of the auxiliary amplifier to the output of the main amplifier will close a feedback loop, which can cause instability. Another problem is that to achieve a large linearity improvement, the accuracy of the summation must be high. The higher the accuracy, the better the distortion cancellation.

For the same reason, the gain of the auxiliary amplifier must be accurate. Feedback can be employed to achieve that. In this way the feedback and feed-forward methods can be combined.

The feed-forward principle has been used in applications with frequencies from audio [6] to RF [7]. The method is particularly attractive at RF, where feedback is not effective due to the limited amount of loop gain available at high frequencies. Instead of requiring the amplifier to be linearized to have high gain, the feed-forward technique requires the additional circuitry to be accurate.

3.3 Predistortion

If a nonlinearity is preceded by a corresponding inverse nonlinearity, the total transfer function will become linear, see figure 3.4a. This is called predistortion. If high linearity is needed over a wide bandwidth, it becomes very difficult to create the inverse nonlinearity, as the nonlinearity of amplifiers tends to be frequency dependent. The simplest example of predistortion is an MOS current mirror, figure 3.4b. Due to internal capacitances, the performance degrades as the frequency is increased [8], that is the linearized bandwidth is small, and in this case centered around DC.

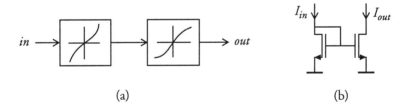

Figure 3.4: (a) The principle of predistortion (b) An MOS current-mirror is a simple example of predistortion

More advanced predistortion topologies are often used at high frequencies [9]. If the predistortion is performed at the intermediate frequency (IF), the power amplifier can be linearized at a frequency band centered around the carrier frequency. To implement the nonlinear functions needed, translinear circuits can, for instance, be used.

An advantage of predistortion compared to feed-forward is that summation at the output is avoided, and an advantage over feedback is that high loop gain is not needed. Like feed-forward, the principle is interesting for RF applications where feedback is not suitable since not enough loop gain can be achieved.

This linearization method requires nonlinear functions to be accurately realized. Furthermore, it requires the nonlinearity of the amplifier to be linearized to be accurately known. To avoid this, some sort of adaptive scheme can be used. Finding the parameter values tends, however, to be complicated, and the result is high complexity.

An alternative approach could be to use the principle sketched in figure 3.5. It uses a model amplifier to find the nonlinearity and create the pre-distorted signal. The model amplifier and its load must be identical to, or a perfectly scaled copy of, the main amplifier with load.

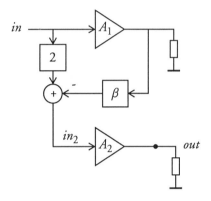

Figure 3.5: Predistortion based on model amplifier

The block (amplifier) with gain equal to two must have high accuracy and low distortion. It should also have the same time delay as A_1. Since it does not have to drive any low-impedance load, it can be implemented as a class A amplifier.

β is to be selected equal to the inverse of the small-signal gain of A_1. Assume the amplifiers to have their gain (normalized) equal to:

$$A_1 = A_2 = 1 + d \tag{3.4}$$

where d is a complex quantity representing the relative distortion. β becomes 1 in this normalized case. The input signal to the second amplifier becomes:

$$in_2 = 2 - (1 + d) = 1 - d \tag{3.5}$$

3.4 Cancellation

The output thus becomes:

$$out = in_2 \cdot A_2 = (1-d) \cdot (1+d) = 1 - d^2 \qquad (3.6)$$

The distortion is thus not completely cancelled using this approach, but it can be largely reduced. If the relative distortion d for instance is 1% (-40dB), d^2 of (3.6) becomes 0.01% (-80dB). The signal to distortion ratio measured in dB is doubled by the squaring of d.

3.4 Cancellation

Cancellation is similar to predistortion. Also here one nonlinearity cancels another. The difference is that the nonlinearities cancel when they are added instead of cascaded, see figure 3.6a. A similarity to feed-forward is the addition of signals at the output.

A simple example of cancellation is an ideal square-law CMOS inverter where the N and P transistors are matched [10], figure 3.6b.

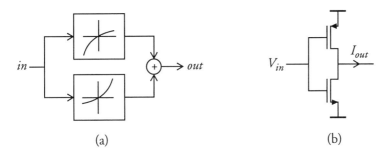

Figure 3.6: (a) The principle of cancellation (b) A CMOS inverter is a simple example of cancellation

Another example is the cancellation of even order nonlinearities in differential amplifier stages and amplifiers. In a differential topology half the input signal and the inverted half input signal are fed to ideally identical nonlinearities. At the output the signals are then subtracted. If the transfer functions of the nonlinearities are static and described by identical polynomials:

$$\begin{cases} Q_{o1} = a_0 + a_1 Q_{i1} + a_2 Q_{i1}^2 + a_3 Q_{i1}^3 + a_4 Q_{i1}^4 + a_5 Q_{i1}^5 \\ Q_{o2} = a_0 + a_1 Q_{i2} + a_2 Q_{i2}^2 + a_3 Q_{i2}^3 + a_4 Q_{i2}^4 + a_5 Q_{i2}^5 \end{cases} \quad (3.7)$$

the output becomes:

$$Q_{out} = Q_{o1} - Q_{o2} = a_0 + a_1 \frac{Q_{in}}{2} + a_2 \left(\frac{Q_{in}}{2}\right)^2$$
$$+ a_3 \left(\frac{Q_{in}}{2}\right)^3 + a_4 \left(\frac{Q_{in}}{2}\right)^4 + a_5 \left(\frac{Q_{in}}{2}\right)^5 - \left\{ a_0 + a_1 \left(-\frac{Q_{in}}{2}\right) \right.$$
$$\left. + a_2 \left(-\frac{Q_{in}}{2}\right)^2 + a_3 \left(-\frac{Q_{in}}{2}\right)^3 + a_4 \left(-\frac{Q_{in}}{2}\right)^4 + a_5 \left(-\frac{Q_{in}}{2}\right)^5 \right\}$$
$$= a_1 Q_{in} + \frac{a_3}{4} Q_{in}^3 + \frac{a_5}{16} Q_{in}^5 \quad (3.8)$$

where it is readily seen that the even order terms have cancelled out each other. If the coefficient pairs of the even order terms are not exactly equal, however, the cancellation will not be perfect.

Other advantages of differential topologies are high immunity to disturbances and doubled available voltage swing. Differential topologies are also well suited for integration as the relative accuracy (matching) of devices on the same chip is excellent. A drawback is the increased complexity, and that the signals might have to be converted to differential at the input or from differential at the output.

3.5 References

[1] H. S. Black, 'Inventing the negative feedback amplifier', IEEE Spectrum, vol. 14, pp. 55-60, Dec. 1977

[2] H. S. Black, 'Wave translation system', U.S. Patent 2,102,671, Dec. 21, 1937

[3] H. W. Bode, *Network Analysis and Feedback Amplifier Design*. D. van Nostrand Company, 1945

[4] H. S. Black, 'Translating system', U.S. Patent 1,686,792, Oct. 9, 1928

[5] J. Vanderkooy and S. P. Lipshitz, 'Feedforward Error Correction in Power Amplifiers', *J. Audio Eng. Soc.*, vol. 28, no. 1/2, pp. 2-16, Jan/Feb 1980

[6] P. J. Walker, 'Current dumping audio amplifier', *Wireless World*, pp. 560-562, Dec. 1975

[7] K. Konstantinou, P. Gardner and D. K. Paul, 'Optimisation Method for Feedforward Linearisation of Power Amplifiers', *Electronics Letters*, vol. 29, no. 18, pp. 1633-1635, Sept. 1993

[8] A. Kristensson, *Design and Analysis of Integrated OTA-C Low-Pass Filters*. Thesis, LUTEDX/(TETE-7081)/1-108(1997)

[9] L. Sundström and M. Johansson, 'Chip for Linearisation of RF power amplifiers using digital predistortion', *Electronics Letters*, vol. 30, no. 14, pp. 1123-1124, July 1994

[10] M. Ismail and T. Fiez (Editors), *Analog VLSI, Signal and Information Processing*, McGraw-Hill, 1994

Chapter 4

Advanced Feedback Techniques

As the description of negative feedback was rather brief in the previous chapter, a more detailed treatment is called for. In this chapter issues such as stability against self oscillation and phase-compensation are therefore dealt with. In the last section of the chapter a novel feedback scheme called feedback boosting is presented. By using this scheme the amount of feedback can become theoretically infinite; that is, the distortion can be completely cancelled.

4.1 The Asymptotic-Gain Model

In practical feedback amplifiers the gain block and the feedback block can be interwoven. If this is the case, the elementary feedback model is no longer suitable, since it clearly requires the gain and feedback blocks to be separate entities. Instead we can use the asymptotic-gain model [1]. Here it is stipulated that an amplifier must use at least one active device, and a model of an active device must contain at least one controlled source. Therefore the amplifier must also contain at least one controlled source. For the purpose of analysis we may denote the controlled source Q_c and the controlling quantity Q_i. Furthermore, let Q_L denote the output quantity and Q_g the driving source. Now, superposition gives [2]:

$$\begin{cases} Q_L = \rho Q_g + \nu Q_c \\ Q_i = \xi Q_g + \beta Q_c \end{cases} \tag{4.1}$$

A non-zero value of β indicates feedback and a non-zero value of ρ indicates a direct path from the input to the output.

Let

$$Q_c = AQ_i \qquad (4.2)$$

The gain of the entire feedback amplifier then becomes:

$$A_f = \frac{Q_L}{Q_g} = \rho + \nu\xi\frac{A}{1-\beta A} \qquad (4.3)$$

As in the elementary feedback model, the product βA is called the loop gain. If the amplifier is ideal, the loop gain is infinite and the gain becomes:

$$A_{f\infty} = \lim_{\beta A \to \infty} A_f = \rho - \frac{\nu\xi}{\beta} \qquad (4.4)$$

This is called the asymptotic gain. Substitution gives:

$$A_f = \frac{Q_L}{Q_g} = \frac{\rho}{1-\beta A} - A_{f\infty}\frac{\beta A}{1-\beta A} \approx A_{f\infty}\frac{-\beta A}{1-\beta A} \qquad (4.5)$$

Note the similarities and differences (no direct term) to (3.1). The sign differences are due to the feedback signal being added to the input signal in (3.1) but subtracted in figure 3.1. If the feedback is to be negative in the asymptotic gain model, either β or A must be negative.

4.2 Stability

The most serious problem associated with feedback is the possibility of instability. Since a part of the output signal is fed back to the input, self-oscillation can occur if the phase shift of the loop gain is not under control.

To determine whether an amplifier with a certain loop gain will be stable or not, the Nyquist diagram can be drawn. In the diagram the complex value of the loop gain is plotted in the s-plane for all positive frequencies, together with its mirror image in the real axis. The amplifier will be stable if and only if the point (1,0) lies outside the enclosed figure [3, 4]. An example of a Nyquist diagram is shown

4.2 Stability

in figure 4.1a, and in figure 4.1b is shown the area around the point (1,0), the gain margin A_m and the phase margin ϕ_m.

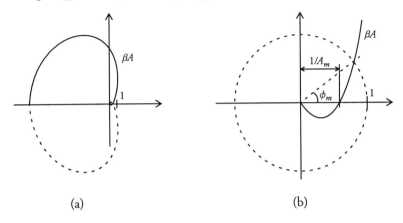

Figure 4.1: (a) Example of Nyquist diagram. (b) Enlarged view around the origin showing gain and phase margins

Due to production tolerances, temperature fluctuations, aging, etc., it is necessary to have some gain and phase margins. Small margins also result in oscillatory step responses.

For design purposes Bode diagrams are preferred over Nyquist. The asymptotic Bode diagrams are particularly useful. In the Bode diagram the phase and the logarithm of the magnitude are plotted versus the logarithm of the frequency. If the Bode diagram is plotted piecewise linear it is called asymptotic. The breakpoints of the magnitude curve are located at the frequencies of the poles and zeros. The corresponding breakpoints of the phase are placed one decade before and one decade after, figure 4.2.

If there are no zeros in the right halfplane (minimum phase), the phase curve can be omitted, since it then can be calculated from the magnitude curve.

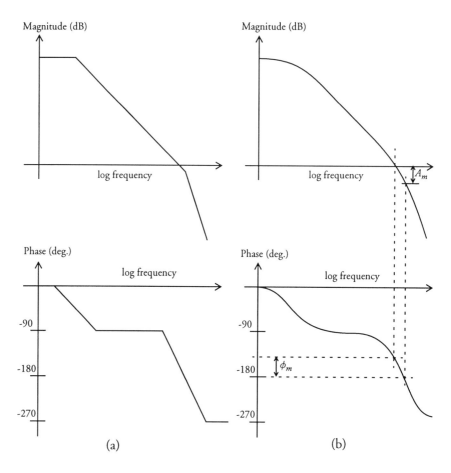

Figure 4.2: Example of Bode diagram. (a) Asymptotic (b) With gain and phase margins

When designing an amplifier it is often better to draw the magnitude curves of A and $1/\beta$ separately, instead of the loop gain. Since the magnitude is drawn logarithmically, the loop gain is then the difference between the two curves. The unity gain frequency is where the curves intersect. If there are no zeros in the right halfplane, the difference of the slopes of the curves at the intersection is important for the stability. If the difference equals or exceeds 40dB per decade, corresponding to a phase shift of 180 degrees or more, the amplifier is either unstable or has too low a phase margin. This observation is very useful when an amplifier is to be phase-compensated.

4.3 Phase-Compensation Techniques

The objective for the phase-compensation is to make the amplifier stable. This is done by shaping the loop gain frequency response so that the phase (and amplitude) margins become positive. To change the loop gain we can change both A and β, as is clear from a Bode diagram with A and $1/\beta$ plotted separately. Many kinds of actions can be taken to alter the frequency responses. Some phase-compensation techniques will result in a significantly lower bandwidth and higher distortion than others. To achieve high performance, good phase-compensation is therefore essential.

The simplest phase-compensation technique is to use a large capacitor to create a dominant pole at a frequency well below any other pole of the amplifier. The dominant pole is used to reduce the loop gain to unity at a sufficiently low frequency, at which the phase shift from the other poles is not large enough to cause instability. This strategy results in poor performance. The bandwidth will be much below optimal, and the high frequency loop gain will be low, resulting in low distortion reduction.

Better, but still not optimal, is to reduce the frequency of the lowest frequency pole, making it dominant. This is called move-pole.

Another possibility is to use internal feedback for phase-compensation. The most common approach is to use a shunt feedback capacitor, also known as Miller compensation. This will result in a pole split, where one pole moves upward in frequency, and one moves downwards. The best is if the compensation can be arranged so that the lowest frequency pole moves down and the second lowest pole moves up. The internal feedback can also help make the amplifier more linear. If all this can be fulfilled, this is an excellent compensation method.

Instead of the forward gain A, the feedback β can be changed. By inserting a left halfplane zero (phantom zero) in β, the phase of the loop gain is advanced. To get the desired effect, the zero is to be placed at a frequency near the unity gain frequency of the feedback loop. If the zero is placed at too low a frequency, the unity gain frequency is increased so much that stability is impaired, due to influence from high frequency poles. Since the feedback network is typically a passive voltage divider, a phantom zero is created by adding a capacitor. This really creates a pole-zero pair and the pole will counteract the zero if they are not well separated. If the zero is placed at too high a frequency, it will not advance the phase sufficiently at the unity gain frequency of the feedback loop. However, correctly executed, this is a very good compensation method [1]. Among the discussed methods, this is the best method, resulting in the largest loop gain, since the high frequency loop gain actually is increased by the compensation.

The methods discussed are illustrated in figure 4.3.

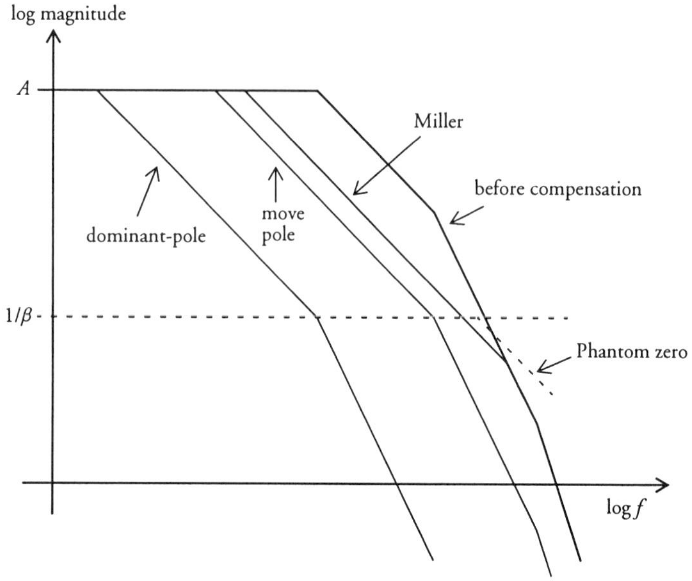

Figure 4.3: Illustration of different phase-compensation methods

As is clear from figure 4.3, the bandwidth and loop gain are different for different compensation methods. What is not shown in the figure is the sensitivity to parameter variations such as production tolerances and different load and source impedances. An advantage of the Miller compensation is that it is very robust against variations [6], but the phantom zero compensation gives maximum distortion reduction [1]. The move pole and dominant pole compensations give inferior performance in terms of loop gain and bandwidth. They also tend to need large capacitors, which is a drawback when the circuit is to be integrated.

In [6] parallel compensation is mentioned. It requires, however, large capacitors and inductors, and is therefore not suitable for integrated circuits. It is also less robust than Miller compensation.

Another option is to use nested feedback loops. An example of this is Miller compensation, where one inner feedback loop is used to phase-compensate an outer loop. The inner loop must also be stable, and if that is achieved using an additional Miller compensation, we have nested Miller compensation [5,6] or nested differentiating feedback loops [7,8].

4.4 Feedback Boosting

If a feedback loop encloses the entire amplifier it is called global, otherwise local. A feedback loop linearizes all stages included in the loop, and global feedback thus linearizes the entire amplifier.

In an amplifier with nested feedback loops, different amplifier stages are linearized by a different number of feedback loops. To achieve the best linearity, the most nonlinear amplifier stage, normally the output stage, should be inside all the feedback loops [7,8]. Such a topology is shown in figure 4.4.

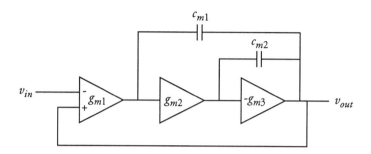

Figure 4.4: A three stage amplifier topology with double nested Miller compensation

Bandwidth requirements can make it necessary to choose another nested Miller compensation where the output stage is not inside all the loops, such as the reversed nested Miller compensation [6].

To achieve high loop gain without using several stages, cascoding is the traditional method. In low voltage amplifiers, this is not possible. Instead, several cascaded amplifier stages must be used. A suitable frequency compensation strategy for multistage amplifiers is to use some sort of nested Miller compensation [6].

4.4 Feedback Boosting

While applying for a patent on improving linearity in amplifiers, I found an already existing patent application presenting a similar topology. This method, which the earlier inventor had called point feedback [9], deserves more recognition. The treatment in this book includes a stability analysis that is vital for the understanding of the principles involved. Also included are measurement and simulation results showing the linearity improvements that can be achieved in a practical circuit.

Complete cancellation of nonlinearity can in theory be accomplished by pre-distortion [10] or feed-forward [11] schemes. In a conventional feedback system a complete cancellation is not possible, due to finite loop gain. A feedback topology that uses infinite loop gain to completely cancel nonlinearity is described here.

Like the other schemes, it is dependent on matching properties. Another drawback is that the infinite loop gain can exist at just one frequency. At other frequencies the loop gain is finite but boosted, making most implementations conditionally stable. The scheme is not as complicated as predistortion, and an advantage over the feed-forward approach is that the problematic summing at the output is avoided.

4.4.1 The Topology

The basic topology of an inverting voltage amplifier with feedback boosting is shown in figure 4.5.

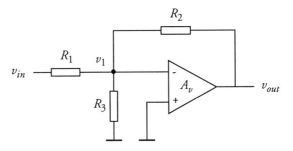

Figure 4.5: The basic inverting topology

Nodal analysis gives:

$$\frac{v_{in}-v_1}{R_1} + \frac{v_{out}-v_1}{R_2} - \frac{v_1}{R_3} = 0 \Rightarrow \frac{v_{out}}{R_2} = -\frac{v_{in}}{R_1} + v_1\left(\frac{1}{R_1}+\frac{1}{R_2}+\frac{1}{R_3}\right)$$
$$= -\frac{v_{in}}{R_1} + \frac{v_{out}}{A_v}\left(\frac{1}{R_1}+\frac{1}{R_2}+\frac{1}{R_3}\right) \quad (4.6)$$

If we select R_3 such that the last term in (4.6) cancels, then the transfer from v_{in} to v_{out} becomes linear and independent of the gain A_v. The gain of the system is

4.4 Feedback Boosting

then completely determined by the resistors R_1 and R_2. The gain is identical to that with an ideal amplifier having infinite gain.

$$R_3 = -(R_1 \| R_2) \qquad (4.7)$$

The last term in (4.6) cancels if R_3 is selected according to (4.7). The negative value does not necessarily result in instability, since the Miller effect of R_2 and the amplifier makes the resistance of the node v_1 positive.

The negative resistance can be implemented as shown in figure 4.6, where it consists of the lower amplifier and R_3, R_4 and R_5. Capacitor C_c is a phase-compensation capacitor that might be necessary depending on the bandwidths of the amplifiers.

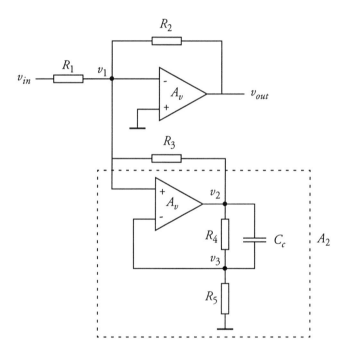

Figure 4.6: The complete inverting topology

4.4.2 Analysis

Assume the lower amplifier with feedback resistors R_4 and R_5 in figure 4.6 to have a frequency response with a single dominant pole at p_1 (4.8).

$$A_2 = \frac{A_{dc2}}{1+s/p_1} \tag{4.8}$$

The negative resistance, R, is then

$$R = -\frac{R_3}{A_2 - 1} = -\frac{R_3}{\frac{A_{dc2}}{1+s/p_1} - 1} = \frac{-R_3(1+s/p_1)}{A_{dc2} - 1 - s/p_1}$$

$$= R_{dc}\frac{1+s/p_1}{1-s/(p_1(A_{dc2}-1))} \quad , \quad R_{dc} = -\frac{R_3}{A_{dc}-1} \tag{4.9}$$

At high frequencies the resistance becomes positive and equal to R_3. This phase-shift is accomplished by the left halfplane zero and the right halfplane pole. The right halfplane pole and negative resistance do not necessarily cause instability, as can be seen by calculating the feedback β of the first amplifier. If R_{dc} is selected according to (4.7), β becomes

$$\beta = \frac{R_1 \| R}{R_2 + R_1 \| R} = \ldots = \frac{R_1}{R_1 + R_2} \cdot \frac{1+s/p_1}{s/p_1} \cdot \frac{A_{dc2}-1}{A_{dc2}} \tag{4.10}$$

The first factor in (4.10) represents the normal feedback, without negative resistance. The rest represents the boost in loop gain. At DC the single s-term in the denominator makes the loop gain infinite. Above p_1 the loop gain is almost unaffected. This pole has to be located well below the frequency where the loop gain reaches one, in order to preserve the phase margin. The whole arrangement results in a conditionally stable amplifier, if the loop gain already rolled off with at least single pole rate in parts of the boosted frequency range.

4.4 Feedback Boosting

If there is a matching error, this will show up as a constant term next to the s in the denominator, reducing the loop gain at low frequencies, see figure 4.7.

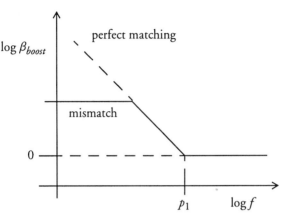

Figure 4.7: The boost in the feedback (extra loop gain)

Even in the case of perfect matching, there will not be zero nonlinearity. The nonlinearity of the second amplifier will cause some errors, but since this amplifier does not have to drive a large load, and since it operates on very small signals (the error signals of the primary amplifier), the nonlinearity will be extremely small.

If we apply some figures we can see how small the signal level at the second amplifier is. Assume the DC-gain to be as low as 1000, and the second amplifier to have the closed-loop gain A_{dc2} equal to 10. The signal at the output of the second amplifier is then 100 times smaller than at the output of the first amplifier. The errors due to mismatch are therefore likely to dominate.

4.4.3 Detailed Stability Analysis

To show the stability of the configuration the characteristic polynomial with single pole amplifiers can be examined.

$$\begin{cases} A_{v1} = \dfrac{A_{o1}}{1 + s/p_1} \\ A_{v2} = \dfrac{A_{o2}}{1 + s/p_2} \end{cases} \tag{4.11}$$

Nodal analysis gives the coefficients of the characteristic polynomial:

$$\begin{aligned} \text{constant term:} \quad & (1 + \beta_2 A_{o2})\left(\frac{1}{R_2} - \frac{1}{A_{o1}}\left(\frac{1}{R_1} + \frac{1}{R_2}\right)\right) + \frac{A_{o2} - 1 - \beta_2 A_{o2}}{R_3 A_{o1}} \\ \text{s-term:} \quad & \frac{1}{p_2}\left(-\frac{1}{R_2} - \frac{1}{A_{o1}}\left(\frac{1}{R_1} + \frac{1}{R_2} + \frac{1}{R_3}\right)\right) + \\ & \frac{1}{p_1} \cdot \frac{A_{o1} - 1 - \beta_2 A_{o2}}{A_{o1} R_3} - \frac{1}{p_1} \cdot \frac{1}{A_{o1}}\left(\frac{1}{R_1} + \frac{1}{R_2}\right) \\ s^2\text{-term:} \quad & \text{negative} \end{aligned} \tag{4.12}$$

β_2 is the feedback of the second amplifier. The configuration is stable if all the coefficients have the same sign. Assuming large A_{o1} and A_{o2}, the stability conditions (negative coefficients) can be simplified:

$$\begin{aligned} \beta_2 R_3 &> \frac{R_2}{A_{o1}} \quad \text{(DC Miller condition)} \\ R_3 &> R_2 \frac{GB_2}{GB_1} \end{aligned} \tag{4.13}$$

The first condition is easy to fulfil. The Gain-Bandwidth product of the second amplifier has to be chosen so that the second condition is fulfilled with some margin.

4.4 Feedback Boosting

4.4.4 Experiment and Simulations

An experiment using two LF351 operational amplifiers was performed. They were used since they have a JFET-input stage with a very high input resistance. One of the amplifiers was heavy loaded and the other was used as a negative resistance.

A distortion meter was used to measure the nonlinearity. The amplifier was phase-compensated not to overshoot on a square wave. The result of the measurement can be seen in figure 4.8. As expected, the feedback-boosting enhances the linearity at low frequencies.

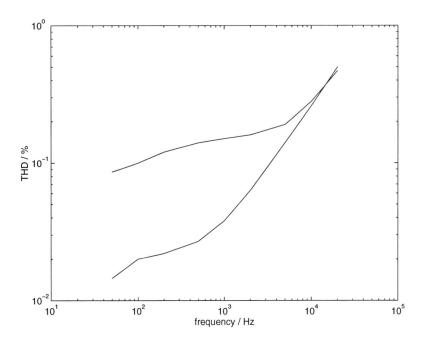

Figure 4.8: Measured THD with and without feedback boosting

The distortion and noise of the instrument made it hard to measure the low distortion at low frequencies. Some simulations were therefore made to verify the low-frequency linearity.

Two integrated two-stage amplifiers in 0.8μm CMOS were used in the simulations. The negative resistance amplifier was made a bit smaller than the other one. The results of the simulations can be seen in figure 4.9. At low frequencies the THD performance was improved several thousand times. Note the correspondence between the distortion reduction in this figure and the feedback boost in figure 4.7 with perfect matching.

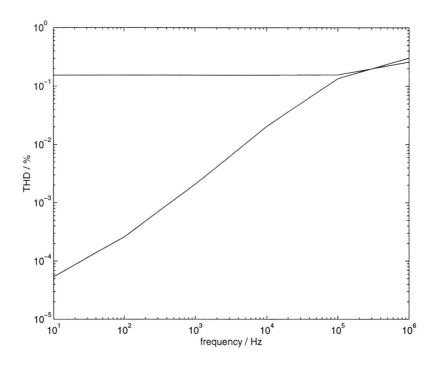

Figure 4.9: Simulated THD with and without feedback boosting (0.8μm CMOS two-stage amplifiers)

4.4.5 Other Topologies

The feedback boosting technique is not limited to inverting voltage amplifiers. A transresistance amplifier can be realized by omitting the resistor R_1 in figure 4.5. Formula (4.7) is then replaced by

$$R_3 = -R_2 \tag{4.14}$$

4.4 Feedback Boosting

It is also possible to realize a non-inverting amplifier with feedback boosting, figure 4.10a. The value of the negative resistance should be selected according to (4.7). The negative resistance has to be floating, which can be realized as in figure 4.10b. The value of the negative resistance is

$$R = -R_4 \frac{R_5}{R_6} \tag{4.15}$$

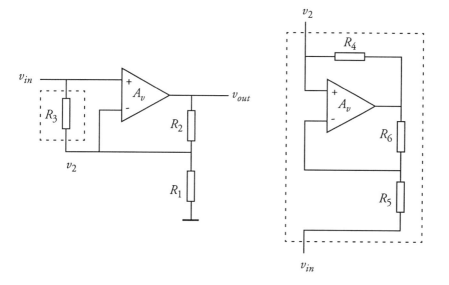

Figure 4.10: (a) A non-inverting topology (b) A floating negative resistance

4.4.6 Distortion Cancellation at Signal Frequencies

The compensation amplifier must have zero phase shift at the frequency where the loop gain is to be infinite. The compensation current can then compensate completely for the error, as the currents are in phase (anti-phase). A compensation amplifier with a low-frequency zero was investigated. The phase-shift was then zero at a non-zero frequency.

Some simulations were made using the same two-stage amplifiers as before. The second amplifier was tuned to have zero phase-shift at 10kHz, using a capacitor in its feedback network. When sending a 5kHz tone through the circuit, the resulting 10kHz (second order) distortion was very small, which shows that the circuit works as intended.

4.5 References

[1] E. H. Nordholt, *Design of High-Performance Negative-Feedbank Amplifiers*. Delftse Uitgevers Maatschappij, 1993

[2] J. Davidse, *Analog Electronic Circuit Design*. Prentice Hall, 1991

[3] H. Nyquist, 'Regeneration Theory', *Bell System Tech. J.*, Jan. 1932, p. 126

[4] P. J. Baxandall, 'Audio power amplifier design - 3', *Wireless World*, pp. 83-88, May 1978

[5] M. J. Fonderie and J. H. Huijsing, *Design of Low-Voltage Bipolar Operational Amplifiers*. Kluwer Academic Publishers, 1993

[6] R. G. H. Eschauzier and J. H. Huijsing, *Frequency Compensation Techniques for Low-Power Operational Amplifiers*. Kluwer Academic Publishers, 1995

[7] E. M. Cherry, 'A New Result in Negative-Feedback Theory, and its Application to Audio Power Amplifiers', Circuit Theory and Applications, vol. 6, pp. 265-288, 1978

[8] E. M. Cherry, 'Nested Differentiating Feedback Loops in Simple Audio Power Amplifiers', *J. Audio Eng. Soc.*, vol. 30, no. 5, pp. 295-305, May 1982

[9] A. M. Sandman, 'Point Feedback', *International Patent Application* WO 97/27668

[10] M. Chadheri, S. Kumar and D. E. Dodds, 'Fast adaptive predistortion lineariser using polynomial functions', *Electronics Letters*, vol. 29, no. 17, pp. 1526-1528, Aug. 1993

[11] J. Vanderkooy and S. P. Lipshitz, 'Feedforward Error Correction in Power Amplifiers', *J. Audio Eng. Soc.*, vol. 28, no. 1/2, pp. 2-16, Jan/Feb 1980

Chapter 5

Output Stages

The task of an output stage is to drive a load, which often requires large output currents. As a result, the output stage often dominates the power consumption of an amplifier.

The modulation, defined as the signal current divided by the quiescent current, is an important figure for the distortion of an amplifier stage. High modulation levels lead to high distortion [1,2]. In a well designed amplifier, the output stage is usually where the signal current and the modulation are largest. From a linearity point of view, the output stage is therefore the most important.

The output stage is thus critical for both power efficiency and linearity. A key decision for both is which class of operation to use. Different classes of operation are discussed in section 5.1. How integrated low voltage output stages can be implemented in both CMOS and bipolar technology is shown in section 5.2.

5.1 Classes of Operation

Normally an amplifier must be able to both source current to and sink current from the load. In other words, some component must be able to carry current from the positive supply to the output, I_p, and some must be able to carry current from the output to the negative supply, I_n, figure 5.1.

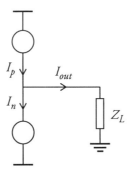

Figure 5.1: The tail currents of the output stage

I_p and I_n are called the tail currents. One or both of the tail currents must be controlled in order to control the output current. The way the tail currents are controlled determines the class of operation.

The power efficiency and linearity are different for different classes of operation. Unfortunately, the best linearity is not achieved for the most power efficient class of operation, so a compromise has to be made. The difference between different classes is well illustrated by plotting the tail currents versus the output current, figure 5.2.

5.1 Classes of Operation

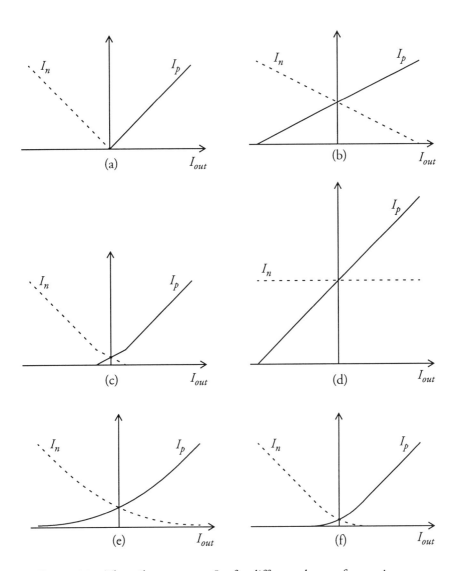

Figure 5.2: The tail currents vs. I_{out} for different classes of operation
(a) class B (b) push-pull class A (c) class AB
(d) current-generator loaded single-ended class A
(e) curved class A (f) curved class AB

The names of the last two classes of operation are my suggestions. They are often referred to as class AB. I think, however, that they differ enough from (c) to have different names. In some cases it can, however, be difficult to tell if an output stage with curved tail currents operates in curved class A or curved class AB.

The intersection between the I_n and I_p curves corresponds to zero output current, and thus gives the quiescent current of the output stage. The class B operation gives zero quiescent current, and has the best power efficiency of the classes illustrated in the figure. For the other classes of operation, the integral of I_p for negative output currents and I_n for positive can be used as a measure of the reduced efficiency compared to class B.

The efficiency of an ideal class B output stage for a sinusoid signal can be calculated as:

$$P_{tot} = \frac{1}{T}\int_0^T (I_n + I_p) V_{sup} \, dt = \frac{1}{\pi}\int_0^\pi \left(\frac{A\sin t}{R_L} V_{sup}\right) dt = \frac{2}{\pi} A \frac{V_{sup}}{R_L}$$

$$P_L = \frac{A^2}{2R_L}$$

$$\eta = \frac{P_L}{P_{tot}} = \frac{\pi}{4} \frac{A}{V_{sup}} \qquad (5.1)$$

where the supply voltage is plus/minus V_{sup}, the amplitude is A and the load resistance is R_L. The efficiency η is proportional to the amplitude, and as the highest possible amplitude before clipping is equal to V_{sup}, the maximum efficiency of a class B stage is $\pi/4 = 78.5\%$.

In a class A stage the total power, P_{tot}, consumed by the amplifier and load is constant. The efficiency then becomes proportional to the power in the load, which is proportional to the square of the signal amplitude:

$$P_{tot} = \begin{cases} I_{max} V_{sup} & \text{(push-pull)} \\ 2 I_{max} V_{sup} & \text{(single-ended)} \end{cases} \qquad (5.2)$$

$$P_L = \frac{A^2}{2R_L} \qquad A < \min(V_{sup}, R_L I_{max}) \qquad (5.3)$$

$$\text{let} \quad x = \frac{R_L I_{max}}{V_{sup}} \qquad (5.4)$$

5.1 Classes of Operation

$$\eta = \frac{P_L}{P_{tot}} = \frac{A^2}{2R_L I_{max} V_{sup}} < \min\left(\frac{x}{2}, \frac{1}{2x}\right) \quad \text{(push-pull)} \quad (5.5)$$

$$\eta = \frac{P_L}{P_{tot}} = \frac{A^2}{4R_L I_{max} V_{sup}} < \min\left(\frac{x}{4}, \frac{1}{4x}\right) \quad \text{(single-ended)} \quad (5.6)$$

where I_{max} is the maximum output current that can be delivered without clipping or leaving class A. The maximum efficiency is achieved for x equal to one. The theoretical limit for the push-pull stage is 50% efficiency, and for the single ended, 25%. As can be seen from the formula, it is important to select the load resistance so that x is near one, in order to avoid an unnecessarily low efficiency.

The quadratic dependence of the efficiency on amplitude for class A results in a poor efficiency for signals with low amplitude. For a class B amplifier the dependence is linear, still resulting in low efficiency for low signals, but much higher than for class A, see figure 5.3.

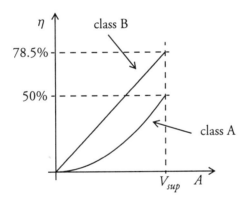

Figure 5.3: Maximum efficiency for class B and push-pull class A vs. amplitude

To achieve an efficiency better than that of class B, pulse-width modulation can be used, class S. In an amplifier tuned to a particular operating frequency, it is possible to produce heavily distorted waveforms that are filtered before they are fed to the load, e.g., classes C, D, E and F. Such amplifiers can have very high efficiency [3].

The sharp bends of the tail currents cause problems. When a signal is amplified they make the required bandwidth of the tail currents infinite. Since the bandwidth is limited, the dynamic tail current diagram for class B will look more like figure 5.2f. Nonlinear effects and delays cause distortion when transistors turn on and off in class B and AB operation. In a class B amplifier this is particularly harmful, since the bend occurs at zero output current. This can result in very low signal to distortion ratios for small signals, and is called cross-over distortion.

In class A operation all devices are conducting all the time. The problem with transistors turning on and off, and thereby cross-over distortion, is avoided. The drawback is the low power efficiency, see figure 5.3. The push-pull class A operation has better efficiency than the single-ended counterpart and often less distortion, as cancellation of even order products often can be accomplished.

The curved class A operation combines the advantages of classes A and B. The quiescent current can be lowered compared to class A, the transistors can be on all the time, and sharp bends can be avoided by making the curves smooth instead of piecewise linear. The distortion performance will generally not be as good as with linear push-pull class A, and the quiescent current is larger than in class B, but for many applications it is an alternative that should be considered.

The curved class AB is an improved class AB operation where the sharp bends have been replaced by smooth curves [4].

5.2   Low Voltage Output Stages

When the supply voltage is low, the output stage has to be designed for maximum output voltage swing. This makes common source (common emitter) coupled output transistors necessary, in order not to loose a threshold voltage (base emitter voltage) at each side [5]. In a bipolar process where pnp transistors can not be used, an output stage with an emitter follower can, however, be the best choice.

In integrated amplifiers, as well as in discrete, output stages working in different classes are used. Usually the rest of the stages operate in class A, as their output currents are small compared to that of the output stage. Class AB stages can, however, be used in order to achieve a high slew-rate. The slew-rate is limited by the maximum current by which an amplifier stage can charge a capacitance, so a large output current capability can thus be needed.

The standard two stage CMOS operational amplifier [6] uses a single-ended current generator loaded class A output stage. The advantage is simplicity, the disadvantage, low power efficiency and reduced output voltage swing in one direction. The distortion before feedback is also larger than for a push-pull class A amplifier. The one stage CMOS operational amplifier [6] uses a push-pull class A output stage.

5.2 Low Voltage Output Stages

In more advanced amplifiers, class AB output stages are often used. The simplest class AB output stage, however, is the ordinary CMOS inverter. More advanced class AB stages have some form of class AB bias control that sets the tail currents.

There are two main categories of class AB bias controls. First are feedback bias controls, where the tail currents are monitored, by for instance a parallel transistor connected to each output device. The sensed bias currents are then used in the bias control that sets the tail currents [7, 8, 9]. The other category is feed-forward bias control. Instead of feedback it can use, for instance, matching to get control of the tail currents. This is the traditional bias method that is used in most class AB amplifiers [10,11-14].

Feedback bias control can be used to create advanced functions for the tail currents versus output current, where all transistors can be prevented from turning off [9]. A problem is, however, that the principle is based on feedback, and phase-compensation thus might be necessary. The bandwidth of the bias control must also greatly exceed that of the signal, as the bending (nonlinearity) of the tail current curves results in increased bandwidth. Furthermore, this type of scheme requires the nonlinearities of the transistors to be well defined – which is not always the case.

Feed-forward techniques are usually simpler, but do not necessarily have inferior performance. This is the type of bias used in the class AB and push-pull class A amplifiers presented later in this book.

The ultimate lower supply voltage limit is $V_{GS}+V_{DSsat}$ for a CMOS amplifier [15] and $V_{BE}+V_{sat}$ for a bipolar. CMOS class AB amplifiers useful at supply voltages down to 1.5V have been reported recently, for instance in [11,15,16]. Bipolar class AB amplifiers have been successfully built for 1V supply [7].

5.2.1 CMOS Output Stages

In order to be able to operate at the minimum supply voltage the output devices must be connected as in figure 5.4. This connection maximizes the output voltage swing. For push-pull operation, such as class B, AB or push-pull class A, V_{Gn} and V_{Gp} must be driven in phase. This can for instance be achieved by a voltage shift between the two inputs. Another approach is to drive the inputs with two separate amplifiers. The amplifiers are in that case designed so that their quiescent output voltages establish the required voltage difference between V_{Gn} and V_{Gp}. Whichever approach is chosen, accurate control of the bias current in the output devices is important.

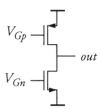

Figure 5.4: The connection of the output devices in a low voltage output stage.

In the approach where two separate amplifiers drive the output devices, the amplifiers can be made identical or complementary, but it is also possible to use completely different amplifiers. The linearity is best if the positive and negative half-periods of the signal are treated similarly; that is, the amplifiers should not have too dissimilar characteristics. Furthermore, it is easier to design a high-performance phase-compensation for an amplifier with similar gain for the positive and negative half-period.

If operation at the minimum supply voltage is required, there may not be any series-connection of more than one gate-source voltage and one drain-source voltage. An amplifier meeting this requirement is called ultimate low voltage [16]. An example of an ultimate low voltage push-pull stage is shown in figure 5.5. A drawback of this scheme is that A_1 and A_2 have very different gain, but an advantage is that the quiescent current is well defined. It is well defined as it relies on matching of the W/L ratios of the transistors:

$$I_{Qout} = I_b \frac{(W/L)_5 (W/L)_1}{(W/L)_4 (W/L)_3} \qquad (5.7)$$

If two drain-source voltages can be allowed in series with one gate-source voltage, the topology of figure 5.6 can be used. The two amplifiers are in this case complementary, making high linearity possible [16].

The output bias current for this topology is:

$$I_{Qout} = I_b \frac{1}{M+1} A_I \qquad (5.8)$$

where M and A_I are accurately determined by matching of transistors.

In [17] a CMOS low voltage amplifier using identical amplifiers (CS-stages) and a feedback class AB control to fix the quiescent current is presented.

5.2 Low Voltage Output Stages

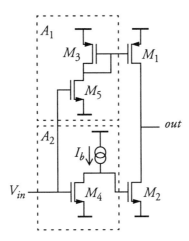

Figure 5.5: Example of ultimate low voltage push-pull output stage with different amplifiers driving the output devices

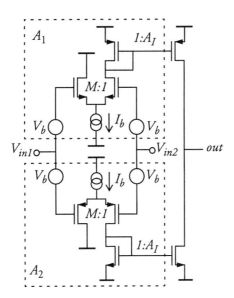

Figure 5.6: Example of low voltage topology with complementary amplifiers

5.2.2 Bipolar Output Stages

For maximum output voltage swing, the output devices in a bipolar amplifier must be connected as in CMOS, figure 5.7.

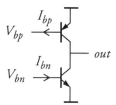

Figure 5.7: Connection of the output devices in a low voltage bipolar amplifier.

As in the CMOS case the two inputs are to be driven in phase to accomplish push-pull operation. An example of how this can be done at the minimum supply voltage is shown in figure 5.8.

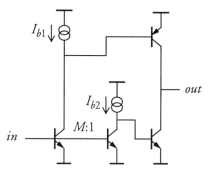

Figure 5.8: Example of low voltage bipolar output stage

The quiescent current is:

$$I_{Qout} = \frac{M \cdot I_{b2} - I_{b1}}{\frac{1}{\beta_{pnp}} + \frac{M}{\beta_{npn}}} \tag{5.9}$$

The quiescent current in this topology is proportional to β and a difference between two quantities, and thereby is not very accurate. A simple modification could be to use a current mirror instead of the current source I_{b1}, as in figure 5.5. More advanced bias circuits based on feedback can be found in for instance [7].

5.2 Low Voltage Output Stages

In some bipolar processes pnp devices show poor performance, or can not be used at all. If the process is a BiCMOS process, PMOS devices can be used instead of pnp, otherwise one has to design the amplifier with just npn transistors. If a push-pull output stage then is to be designed, npn devices must be used for both push and pull, see figure 5.9.

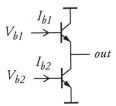

Figure 5.9: Connection of output devices in an npn-only bipolar process.

The drawback is that the output voltage swing is reduced by one base-emitter voltage compared to figure 5.7. The inputs are to be driven out of phase for push-pull operation. The voltage swing requirements are larger at the input of the upper device than at the lower, since it is connected as an emitter follower. An example of an npn-only push-pull output stage is shown in figure 5.10.

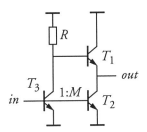

Figure 5.10: Example of npn-only push-pull stage

The quiescent current is:

$$I_{Qout} = \frac{M\beta}{M+\beta} \frac{V_{CC} - V_{OUT} - V_{BE1}}{R} \approx M \frac{V_{CC} - V_{BE1}}{2R} \quad (5.10)$$

where M can be made very accurate using matching. The accuracy of the quiescent current is thus determined by the accuracy of the resistance and the supply voltage.

5.3 References

[1] P. J. Baxandall, 'Audio power amplifier design - 5', *Wireless World*, pp. 53-56, Dec. 1978

[2] P. J. Baxandall, 'Audio power amplifier design - 6', *Wireless World*, pp. 69-73, Feb. 1979

[3] H. L. Krauss, C. W. Bostian and F. H. Raab, *Solid State Radio Engineering*. Wiley, 1980

[4] I. Hegglun, 'Square law rules in Audio Power', *Electronics World + Wireless World*, pp. 751-756, Sept. 1995

[5] R. Hogervorst, J. P. Tero, R. G. H. Eschauzier and J. H. Huijsing, 'A Compact Power-Efficient 3 V CMOS Rail-to-Rail Input/Output Operational Amplifier for VLSI Cell Libraries', *IEEE Journal of Solid-State Circuits*, vol. 29, no. 12, pp. 1505-1512, Dec. 1994

[6] R. Gregorian and G. C. Themes, *Analog MOS Integrated Circuits for Signal Processing*. Wiley, 1986

[7] M. J. Fonderie and J. H. Huijsing, *Design of Low-Voltage Bipolar Operational Amplifiers*. Kluwer Academic Publishers, 1993

[8] E. Seevinck, W. de Jager and P. Buitendijk, 'A Low-Distortion Output Stage with Improved Stability for Monolithic Power Amplifiers', *IEEE Journal of Solid-State Circuits*, vol. 23, no. 3, pp. 794-801, June 1988

[9] J. H. Botma, R. J. Wiegerink, S. L. J. Gierkink and R. F. Wassenaar, 'Rail-to-Rail Constant-G_m Input Stage and Class AB Output Stage for Low-Voltage CMOS Op Amps', *Analog Integrated Circuits and Signal Processing*, vol. 6, no. 2, pp. 121-133, Sept. 1994

[10] R. Hogervorst, J. P. Tero, R. G. H. Eschauzier and J. H. Huijsing, 'A Compact Power-Efficient 3 V CMOS Rail-to-Rail Input/Output Operational Amplifier for VLSI Cell Libraries', *IEEE Journal of Solid-State Circuits*, vol. 29, no. 12, pp. 1505-1512, Dec. 1994

[11] R. van Dongen and V. Rikkink, 'A 1.5 V Class AB CMOS Buffer Amplifier for Driving Low-Resistance Loads', *IEEE Journal of Solid-State Circuits*, vol. 30, no. 12, pp. 1333-1337, Dec. 1995

[12] R. G. Meyer and W. D. Mack, 'A Wide-Band Class AB Monolithic Power Amplifier', *IEEE Journal of Solid-State Circuits*, vol. 24, no. 1, pp. 7-11, Feb. 1989

5.3 References

[13] K. Nagaraj, 'Large-Swing CMOS Buffer Amplifier', *IEEE Journal of Solid-State Circuits*, vol. 24, no. 1, pp. 181-183, Feb. 1989

[14] J. A. Fisher and R. Koch, 'A Highly Linear CMOS Buffer Amplifier', *IEEE Journal of Solid-State Circuits*, vol. SC-22, no. 3, pp. 330-334, June 1987

[15] R. G. H. Eschauzier, R. Hogervorst and J. H. Huijsing, 'A Programmable 1.5 V CMOS Class-AB Operational Amplifier with Hybrid Nested Miller Compensation for 120 dB Gain and 6MHz UGF', *IEEE Trans. of Solid-State Circuits*, vol. 29, no. 12, pp. 1497-1504, Dec. 1994

[16] H. Sjöland, 'A 1.5V Class AB CMOS Power Amplifier', *Proc. NorChip '97*, pp.180-186, Tallinn, Estonia, Nov. 1997

[17] R.G.H. Eschauzier, J.H.Huijsing, *Frequency Compensation Techniques for Low-Power Operational Amplifiers*, Kluwer Academic Publishers, 1995

Chapter 6

Analysis and Measurement of Distortion

This chapter deals with simulation and measurement of distortion. The methods use input signals consisting of one or more sinusoids. They can not handle a wideband input signal. Therefore a novel method for relating intermodulation distortion of a wideband signal to harmonic distortion of a sinusoid is presented. The harmonic distortion can then be simulated or measured, and from that the intermodulation distortion can be estimated. This method has also been presented in IEEE Transactions on Circuits and Systems [1,2], and I am grateful to IEEE for letting me use the material for this chapter.

6.1 Computer Simulation of Distortion

Before an integrated circuit design is sent for fabrication it must be verified by means of simulation. This is necessary, since once a circuit has been fabricated, the only way to make changes is by refabrication. If the circuit has been properly simulated, the probability of a working circuit is increased. The simulations give an indication of what performance to expect. Different types of simulations have to be performed to find the different amplifier figures of merit discussed in section 2.2.

There are different methods for simulating distortion [3]. The most straightforward, and probably most accurate, is to perform a transient analysis, and analyse the result using a Fast Fourier Transform (FFT). In the transient analysis the nonlinear differential equations of the circuit are solved numerically in the time-domain. The FFT is used to transform the result to the frequency domain, where the amplitudes of the signal components can be found. The THD, IP_3 and so on, can then be calculated. Both single-tone and multi-tone tests can be performed using this method. The main drawback is that for complicated circuits the simulation time can be long. Multi-tone tests generally require more time than single-tone tests, as the intermodulation products can have frequencies much lower than the test tones, and the simulation time is determined by the lowest frequency component.

An extra time interval has to be simulated at the start, in order for start-up transients to die out. That time interval is not to be included in the FFT.

The accuracy of a simulation result is never better than that of the device models used. This is especially harmful for distortion simulation, as it is hard to accurately model the nonlinearity of the devices. One reason why a transistor is hard to model is that at different regions of operation the behaviour can be quite different (e.g., weak inversion and strong inversion in MOS). The model must be accurate in the different regions, as well as in the transitions between them [4].

6.2 Distortion Measurement

Depending on the frequency, the distortion is measured using different equipment. At audio frequencies THD meters are available. They measure the output and present the THD-figure directly. A problem with THD meters is that the noise inside the measurement bandwidth usually is added to the measured THD figure. Most THD meters contain signal generators, and more advanced instruments allow sweeping to facilitate THD versus, for instance, frequency measurements. There are also audio analysers available that together with a computer create quite advanced measurement systems. The computer is used for calculations and presentation of results.

There are spectrum analysers available for frequencies from audio to RF. The amplifier to be measured is fed with one or two tones using signal generators, and the output is connected to the spectrum analyser. The spectrum analyser presents the signal in the frequency domain. The frequency components can then be measured and the distortion calculated.

If the amplifier to be measured is very linear, it is necessary to be careful not to measure the distortion of the measurement equipment rather than the amplifier. In a single-tone measurement, it might, for instance, be necessary to precede the amplifier by a low-pass filter, to suppress harmonics from the signal generator.

6.3 Intermodulation Distortion related to THD

The previous sections described how distortion can be simulated and measured with methods using single-tone and multi-tone input signals. Signals in real applications, however, rarely consist of one or a number of tones. Instead, they contain information at a band of frequencies. One way of dealing with this is to relate the distortion of the real signal to the distortion of a signal consisting of tones. The distortion of the real signal can then be estimated from a measurement or simulation using tones.

A novel method that relates the distortion of a wideband signal to that of a single tone (THD) was therefore developed [1,2]. Statistics were used to find which amplitude and frequency a single-tone input signal must have in order to cause the same amount of distortion as a wideband signal. Both static and dynamic nonlinearity can be handled. The static nonlinearity determines the amplitude of the tone, and the dynamic nonlinearity determines the frequency. The method is described in the subsequent sections.

A method based on the summing of several sinusoids to form the wideband signal has been presented in [5]. This method did not, however, include dynamic nonlinearity. Clipping was not considered either, and the results are therefore very pessimistic for high order distortion. The reason is that there is no amplitude limit for a wideband signal, though the probability of extreme amplitudes is low. If these large amplitudes are fed to a high order polynomial, associated with high order distortion, the distortion can become enormous. In reality the distortion is limited in those cases by amplifier clipping. For low order nonlinearities the results of the two methods correspond.

6.3.1 Models of the Nonlinear Amplifier

In order to perform a mathematical analysis of the distortion, the amplifier and its nonlinearity must be described by a mathematical model. The model is preferred to be simple to simplify the analysis, but must still be general enough to cover most cases of interest. This work deals with wideband amplifiers, so we want to find a simple model that is accurate enough for most wideband amplifiers.

Let's start with the static nonlinearity. A simple but general model for the static nonlinearity is shown in figure 6.1:

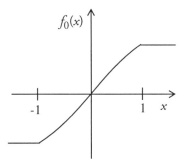

Figure 6.1: Static characteristic

The characteristic is normalized so that the maximum input amplitude before clipping and the gain are equal to one. The nonlinearity before clipping is assumed to be soft and is therefore suitable for description by a polynomial:

$$f_0(x) = \begin{cases} f_0(1) & x > 1 \\ x + a_2 x^2 + a_3 x^3 + \ldots & |x| \leq 1 \\ f_0(-1) & x < -1 \end{cases} \quad (6.1)$$

The error (nonlinearity) before clipping becomes:

$$g_0(x) = a_2 x^2 + a_3 x^3 + \ldots \quad (6.2)$$

The dynamic nonlinearity is modelled similar to the static. As will be seen later, this will simplify matters as the dynamic and static nonlinearity can be handled similarly by this method. To model the dynamic nonlinearities, the time derivatives of the input signal are considered. Just as in the static case, clipping can occur due to large signals. For the first order derivative this is identified as slew-rate clipping.

As opposed to static distortion, it is not possible to calculate the amount of dynamic distortion by just knowing the present input signal. Instead, all the past values of the input signal, the history, must be known. However, as a statistical

6.3 Intermodulation Distortion related to THD

method will be used, it is sufficient to know the average amount of distortion as a function of the input signal. The average distortion contribution can be written as a function of the present input signal. To simplify further, the distortion contributions of the different time derivatives are assumed to be independent. The total average distortion power is modelled as:

$$\overline{P_d} = [f_0(x)]^2 + [f_1(x')]^2 + [f_2(x'')]^2 + [f_3(x^{(3)})]^2 + \ldots \qquad (6.3)$$

where $f_0(x)$ models the static nonlinearity. All f_n-functions give mean errors and are assumed to be soft before clipping.

The assumption that the distortion contributions from different derivatives are independent is reasonable for most wideband amplifiers. An amplifier relying on cancellation between derivatives can not be well modelled by (6.3), but the required balance conditions in such an amplifier are typically narrow-band, and not used in wideband amplifiers.

6.3.2 Input Signals

To make the comparison of the distortion of a sinusoid and a wideband signal, it is necessary to have suitable descriptions of the signals. The statistical method uses the probability density functions [6] of the input signals and their derivatives.

Let's start with the sinusoid:

$$x = A\sin(2\pi f_s \cdot t) \qquad (6.4)$$

The probability density function (pdf) is given by:

$$p_s(x) = \begin{cases} \frac{1}{\pi} \cdot \frac{d}{dx}\mathrm{asin}\left(\frac{x}{A}\right) = \frac{1}{\pi A} \cdot \frac{1}{\sqrt{1-\left(\frac{x}{A}\right)^2}} & |x| < A \\ 0 & |x| > A \end{cases} \qquad (6.5)$$

A plot of the function is shown in figure 6.2:

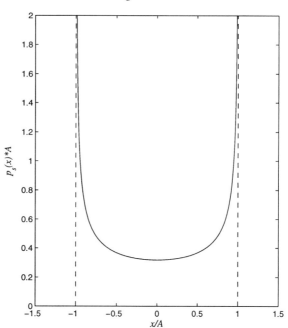

Figure 6.2: The probability density function of a sinusoid with amplitude A

The pdf of a sinusoid is independent of phase and frequency; only the amplitude is important. The derivatives of a sinusoid are also sinusoids, but with different phases and amplitudes. The pdf of the nth derivative is found by replacing A in (6.5) with A_n of (6.6).

$$A_n = A \cdot (2\pi f_s)^n \tag{6.6}$$

Now we move on to the wideband signal. We assume it to be Gaussian; that is, the pdf is given by:

$$p_g(x) = \frac{1}{\sigma\sqrt{2\pi}} \cdot e^{-\frac{1}{2}\cdot\left(\frac{x}{\sigma}\right)^2} \tag{6.7}$$

If the signal consists of a number of independent signals, such as a number of radio channels, it is a good approximation to assume it to be Gaussian. Noise is also often assumed to be Gaussian, so it is reasonable to assume the wideband

6.3 Intermodulation Distortion related to THD

signal to be Gaussian. A plot of the pdf is shown in figure 6.3. The non-zero value outside the dotted lines will cause clipping distortion. It will be lower if σ is decreased, but will never disappear completely.

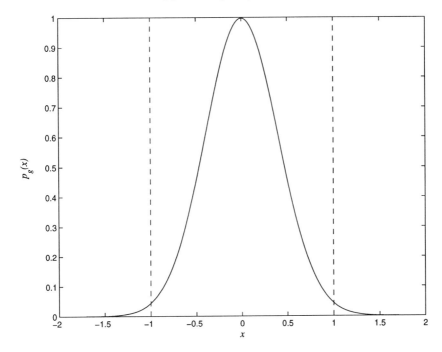

Figure 6.3: The probability density function of the Gaussian wideband signal with $\sigma=0.4$. The dashed lines indicate clipping.

To describe the signal it is also necessary to specify its frequency content. The wideband signal is assumed to have constant spectral density $R_x(f)$ from 0 up to f_w, and zero above f_w, see figure 6.4.

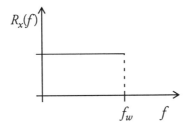

Figure 6.4: The spectral density of the wideband signal

A very useful property of a Gaussian signal is that the *n*th order derivative of the signal also is Gaussian. The spectral density, however, is no longer flat. Instead it is given by:

$$R_{x^{(n)}}(f) = (2\pi f)^{2n} \cdot R_x(f) \tag{6.8}$$

It is possible to calculate the variance of the derivatives by integrating the spectral densities:

$$\sigma_{x^{(n)}}^2 = \int_0^{f_w} (2\pi f)^{2n} R_x(f) df = \frac{\sigma_x^2}{f_w}(2\pi)^{2n} \cdot \frac{f_w^{2n+1}}{2n+1} = \sigma_x^2 \cdot \frac{(2\pi f_w)^{2n}}{2n+1} \tag{6.9}$$

Now the pdf's of both the sinusoid and the wideband signal with their derivatives have been found, which is sufficient for the method to be used.

6.3.3 Static Distortion

The method starts by comparing the intermodulation distortion and THD for a static nonlinearity, which is simpler than a dynamic. As a result, a suitable amplitude for the sinusoid is found. The method is then augmented to handle dynamic nonlinearities, and the frequency of the sinusoid can then also be determined.

6.3.3.1 Static Clipping

The static clipping is caused by the input signal having an amplitude so large that the amplifier enters clipping. In the normalized case of figure 6.1 this corresponds to a magnitude larger than one. If the probability distribution of the input signal is known, the amount of clipping distortion can be calculated. For a Gaussian distribution, the (static) clipping distortion power becomes:

$$P_{clip} = \int_{-\infty}^{-1} (x - f(-1))^2 p_g(x) dx + \int_1^{\infty} (x - f(1))^2 p_g(x) dx \approx$$

$$2\int_1^{\infty} (x-1)^2 p_g(x) dx = \frac{2}{\sigma\sqrt{2\pi}} \int_1^{\infty} (x^2 - 2x + 1) e^{-x^2/(2\sigma^2)} dx \tag{6.10}$$

6.3 Intermodulation Distortion related to THD

The integral was evaluated numerically for different values of σ, and related to maximum sinusoid power before clipping, with the following results:

$$\sigma = \frac{1}{4} \Rightarrow P_{clip} \approx -61\,\text{dB}$$

$$\sigma = \frac{1}{5} \Rightarrow P_{clip} \approx -85\,\text{dB}$$

$$\sigma = \frac{1}{6} \Rightarrow P_{clip} \approx -113\,\text{dB}$$

$$\sigma = \frac{1}{7} \Rightarrow P_{clip} \approx -144\,\text{dB}$$

The lower σ is, the lower the probability is of large signal amplitudes resulting in clipping, and thereby also the power of clipping distortion. Depending on the distortion requirements, a sufficiently low variance (σ) must be chosen. The variance must not be selected too low, however, since the signal power and thereby the signal to noise ratio then will suffer.

6.3.3.2 Static Distortion Before Clipping

According to section 6.3.1 the nonlinearity of the characteristic before clipping is soft and can be described by a polynomial $g_0(x)$. The static distortion power due to different x^n-terms of the polynomial is calculated for both the input signals. In these calculations the pdf's of the input signals are used.

We start with the distortion of a sinusoid due to x^n using [7, formula 82, page 146] to solve the integral:

$$P_{S_n} = \int_{-A}^{A} x^{2n} p_s(x)\,dx = \frac{1}{\pi A} \int_{-A}^{A} \frac{x^{2n}}{\sqrt{1 - \left(\frac{x}{A}\right)^2}}\,dx = \left\{ t = \frac{x}{A} \right\}$$

$$= A^{2n} \cdot \frac{1}{\pi} \int_{-1}^{1} \frac{t^{2n}}{\sqrt{1-t^2}}\,dt = A^{2(n-1)} \cdot A^2 \cdot \frac{1}{\pi} \cdot \frac{2n-1}{2n} \int_{-1}^{1} \frac{t^{2(n-1)}}{\sqrt{1-t^2}}\,dt$$

$$= A^2 \cdot \frac{2n-1}{2n} \cdot P_{S_{n-1}} \tag{6.11}$$

$$P_{S_0} = \int_{-A}^{A} p_s(x)\,dx = 1 \tag{6.12}$$

Recursion gives:

$$P_{S_n} = A^{2n} \cdot \frac{2n-1}{2n} \cdot \frac{2n-3}{2n-2} \cdot \frac{2n-5}{2n-4} \cdot \ldots \cdot \frac{1}{2} \qquad (6.13)$$

In a THD test not all of P_{Sn} will result in distortion, because some of the power will be at the fundamental frequency or at DC, see table 6.1. The ratio of the harmonic power to total power is referred to as harmonic content $H(n)$.

Table 6.1 : Distribution of distortion caused by different orders of nonlinearity with a sinusoid input signal.

order (n)	trigonometric expression [7]	harmonic power $H(n)$	fundamental power	DC power
2	$\sin^2 t = -0.5\,(\cos 2t - 1)$	50%	0	50%
3	$\sin^3 t = -0.25\,(\sin 3t - 3\sin t)$	10%	90%	0
4	$\sin^4 t = 1/8\,(\cos 4t - 4\cos 2t + 3)$	65%	0	35%
5	$\sin^5 t = 1/16\,(\sin 5t - 5\sin 3t + 10\sin t)$	20%	80%	0

For the Gaussian input signal the distortion before clipping becomes:

$$P_{G_n} = \int_{-1}^{1} x^{2n} p_g(x)\,dx = \frac{1}{\sigma\sqrt{2\pi}} \int_{-1}^{1} x^{2n} e^{-\frac{x^2}{2\sigma^2}}\,dx \qquad (6.14)$$

It is not possible to solve this integral analytically, but if the boundaries are removed so that the integration is performed over an infinite interval, it can be solved [7, formula 42, p. 164].

$$P_{G_n(unbound)} = \int_{-\infty}^{\infty} x^{2n} e^{-\frac{x^2}{2\sigma^2}}\,dx = \ldots = (2n-1)(2n-3) \cdot \ldots \cdot 1 \cdot \sigma^{2n} \qquad (6.15)$$

Now enough equations are found for calculating the THD/IM ratio. For small values of σ and n, the $P_{Gn(unbound)}$ can be used, resulting in:

$$\left(\frac{\text{THD}}{\text{IM}}\right)_n = \frac{P_{S_n} \cdot H(n)}{P_{G_n(unbound)}} = \left(\frac{A}{\sigma}\right)^{2n} \cdot \frac{1}{2^n \cdot n!} \cdot H(n) \qquad (6.16)$$

6.3 Intermodulation Distortion related to THD

To be able to estimate the intermodulation (IM) of a wideband signal from a THD-test, we want to find an amplitude A that makes the THD as equal to the IM as possible. The A/σ-ratio should thus be selected so that the THD/IM-ratio becomes as close to one as possible. Using (6.16) the THD/IM-ratio is plotted versus A/σ for different orders of nonlinearity (n) in figure 6.5.

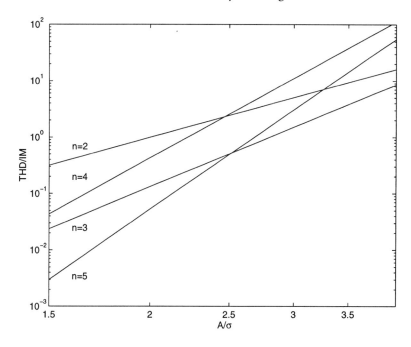

Figure 6.5: THD/IM vs. A/σ and n

As can be seen in the figure an A/σ-ratio of 2.5 is a good choice. If odd orders dominate, a slightly higher ratio of about 2.7 is more suitable, and if even orders dominate, a ratio of about 2.2 is best. This is due to the lower harmonic content $H(n)$ associated with the odd orders in a THD-test.

In order to get analytical expressions, the above results were derived without regarding amplifier clipping. To investigate the effect of clipping, numerical calculations were used to plot the THD/IM versus A and n when the A/σ ratio was fixed to 2.5, see figure 6.6.

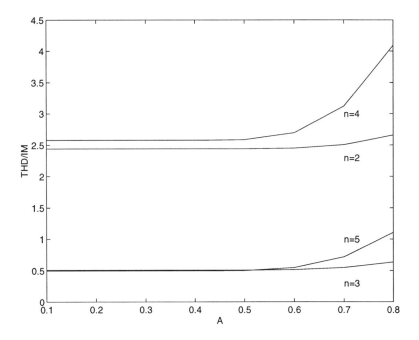

Figure 6.6: THD/IM vs. A and n when $A/\sigma = 2.5$

The bend upward at large amplitudes is due to the influence from the clipping limits. By inspection of the figure it is clear that for n up to 5 and A below about 0.6 there is no significant influence of the clipping limits, so the analytical result can then be used without adjustments.

6.3.4 Dynamic Distortion

The static distortion has now been treated, and we move on to the more complicated dynamic distortion. The static and dynamic nonlinearities are treated equally in equation (6.3) modelling the nonlinear amplifier. This, combined with the similarity of the pdf's of the input signals and their respective derivatives, allows the results from the static distortion to be used also for the dynamic case.

6.3 Intermodulation Distortion related to THD

6.3.4.1 Slew-Rate Clipping

Just as for the static distortion, clipping is handled separately. In the dynamic case clipping occurs due to large time derivatives of the signal. The first order derivative clipping, known as slew-rate clipping, is usually dominant, and we therefore concentrate on that.

The time derivative of the output can not exceed the slew-rate (SR). The slew-rate is assumed to be equal for positive and negative derivatives. If it is not equal, the lowest value will be referred to as SR. If the desired output derivative exceeds SR, a gross distortion called slew-rate clipping occurs [8].

Let d be the derivative of the input signal, and D the derivative normalized with SR:

$$d = \dot{x} \qquad D = \frac{\dot{x}}{SR} \qquad (6.17)$$

The slew-rate clipping distortion power can then be written as:

$$P_{SRtot} = \int_{-\infty}^{-1} p(D)P_{SR}(D)dD + \int_{1}^{\infty} p(D)P_{SR}(D)dD = \left\{ \begin{array}{l} \text{Gaussian} \\ \text{Symmetry} \end{array} \right\}$$

$$= 2\int_{1}^{\infty} \frac{1}{\sigma_D \sqrt{2\pi}} e^{-\frac{D^2}{2\sigma_D^2}} \cdot P_{SR}(D)dD \qquad (6.18)$$

where $P_{SR}(D)$ is the average clipping distortion power as a function of D. There is a similarity between (6.18) and the corresponding formula for static clipping (6.10). In both equations the order of magnitude is determined by the exponential function. A consequence is that if the function $P_{SR}(D)$ does not differ a lot from the polynomial of (6.10), σ_D is to be selected approximately equal to σ for the static and slew-rate clipping to be equal in magnitude. Combined with (6.9) this gives:

$$\left. \begin{array}{l} \sigma \approx \sigma_D \\ \sigma_D \cdot SR = \sigma \cdot 2\pi \cdot \frac{f_w}{\sqrt{3}} \end{array} \right\} \Rightarrow SR \approx 2\pi \cdot \frac{f_w}{\sqrt{3}} \qquad (6.19)$$

This equation approximately gives the minimum slew-rate required from a wide-band amplifier fed by a Gaussian input signal with constant spectral density. The amplifier must be able to produce a sinusoid output signal with the maximum amplitude at $f_w/\sqrt{3}$ without slew-rate clipping. This requirement is independent of the required distortion performance, as the signal level is selected low enough to fulfil the requirements of static clipping. The slew-rate clipping then also becomes sufficiently low, since it is approximately equal to the static clipping when (6.19) is fulfilled.

It must be shown that $P_{SR}(D)$ behaves as stated (like the polynomial of (6.10)), but as the exponential function decays so rapidly with σ_D in (6.18), a very approximate estimate of $P_{SR}(D)$ is sufficient. The sketch of a typical slew-rate clipping scenario shown in figure 6.7 will aid us putting up the equations.

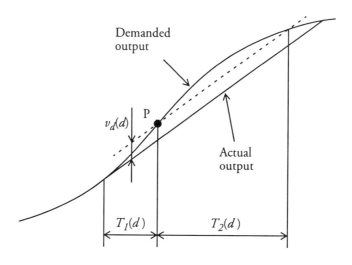

Figure 6.7: Slew-rate clipping scenario

From the figure it is clear that slew-rate clipping will result in an error at the output during a time interval following the causing signal. To handle this, the additional distortion energy during the time interval is calculated and assigned to the causing signal. It is then possible to calculate the power $P_{SR}(D)$. This is an average quantity, as the slew-rate clipping is a complicated process that never looks the same.

During the time $T_2(d)$ the distortion voltage $v_d(d)$ causes the distortion energy:

$$E_{d1} = T_2(d) \cdot [v_d(d)]^2 \tag{6.20}$$

6.3 Intermodulation Distortion related to THD

If an additional distortion voltage v_{add} is added, the energy instead becomes:

$$E_{d2} = T_2(d)[v_d(d) + v_{add}]^2 = T_2(d)[v_d(d)^2 + 2v_d(d)v_{add}^2 + v_{add}^2]$$
$$= E_{d1} + T_2(d)[2v_d(d)v_{add} + v_{add}^2] \qquad (6.21)$$

The additional energy due to v_{add} thus becomes:

$$E_{add} = T_2(d)[2v_d(d)v_{add} + v_{add}^2] \qquad (6.22)$$

In figure 6.7 the additional voltage v_{add} due to the slew-rate in a small time interval ΔT centered around the point P is:

$$v_{add}(d) = (d - SR)\Delta T \qquad (6.23)$$

The additional energy then becomes:

$$E_{add}(d) = T_2(d) \cdot \{[(d - SR)\Delta T]^2 + 2v_d(d)(d - SR)\Delta T\} \qquad (6.24)$$

The power $P_{SR}(D)$ can written as:

$$P_{SR}(D) = \lim_{\Delta T \to 0} \frac{E_{add}(d)}{\Delta T} = T_2(d) \cdot 2v_d(d)(d - SR) \qquad (6.25)$$

An approximation for $v_d(d)$ is made:

$$v_d(d) = (d - SR)T_1(d) \qquad (6.26)$$

Combined with (6.25) this gives:

$$P_{SR}(D) = 2(d - SR)^2 T_1(d) T_2(d) = 2(D - 1)^2 SR^2 T_1(D) T_2(D) \qquad (6.27)$$

When written in this form, the only difference from the polynomial in (6.10) is the factor $2SR^2 T_1(D)T_2(D)$ that always is between zero and two. The factor is usually less than one, so that the slew-rate clipping is slightly over-estimated, but as mentioned earlier, the accuracy requirement is relaxed by the exponential function in (6.18).

6.3.4.2 Dynamic Distortion Before Clipping

The amplitude of the sinusoid in the THD test is determined by the static non-linearity. We now want to find the frequency as well. The frequency f_n, where the distortion due to the nth derivative in (6.3) is similar for the sinusoid and the wideband signal, is sought. As the derivatives of sinusoids and Gaussians also are sinusoids and Gaussians, it is possible to use the result from the static distortion before clipping, which relates the distortion caused by a sinusoid to that of a Gaussian.

For the static case, the amplitude is to be selected as:

$$A = k \cdot \sigma \qquad (6.28)$$

where k is equal to 2.5 when there is balance between even and odd order distortion, slightly larger when the odd orders dominate, and correspondingly slightly smaller if the even orders are the strongest.

For the dynamic case this corresponds to:

$$A_n = k \cdot \sigma_{x^{(n)}} \qquad (6.29)$$

Using (6.6) and (6.9) this can be written as:

$$A \cdot (2\pi f_n)^n = k \cdot \sigma \cdot \frac{(2\pi f_w)^n}{\sqrt{2n+1}} = \{A = k \cdot \sigma\} = A \cdot \frac{(2\pi f_w)^n}{\sqrt{2n+1}} \qquad (6.30)$$

f_n can then be expressed as:

$$f_n = \frac{f_w}{\sqrt[2n]{2n+1}} \qquad (6.31)$$

This equation tells us how to select the frequency of the test tone, depending on which order n of the derivatives dominates.

6.3.5 The Method

The result of all the analysis can now be summarized in a method of how to relate the intermodulation distortion of a wideband signal to the THD of a sinusoid.

The first step of the method is to determine an appropriate variance (power) of the wideband signal. This is done based on static clipping, see section 6.3.3.1. Depending on how much clipping distortion can be accepted, a sufficiently low variance is chosen. The variance should not be selected unnecessarily low, however, as the signal to noise ratio then would suffer.

When the variance σ has been chosen, the amplitude of the sinusoid is set to 2.5σ, which is determined by the static nonlinearity before clipping. If the nonlinearity mainly is of odd order, the amplitude should be somewhat larger, and correspondingly lower if the even orders dominate, see section 6.3.3.

The frequency of the sinusoid can now be determined using (6.31). If the dynamic distortion is dominated by the first order derivative, the frequency is to be selected as $f_w/\sqrt{3}$. At the same frequency ($f_w/\sqrt{3}$), the amplifier must be able to produce a sinusoid of full amplitude without slew-rate clipping.

If the amplitude and frequency are selected in this way, the distortion power will be similar for the two input signals. The intermodulation distortion can thereby be estimated by a THD test.

6.3.6 Example

A wideband IF amplifier is used as an example. The total noise in the operating band is 60dB below maximum sinusoid power before clipping, and we want the intermodulation power to be below this.

The clipping limit at the input, v_{inmax}, is 17.5mV. We assume the variance σ to be equal to $v_{inmax} \cdot 1/4$, resulting in -61dB clipping distortion. The next step is to choose the amplitude A for the sinusoid. In this example we have a differential topology so that the odd orders dominate. We therefore choose $k=2.7$, resulting in $A=11.8$mV.

The first order derivative dominates the dynamic distortion, so we choose the highest operating frequency divided by the square root of three as the frequency of the test tone. In this case $f_w=20$MHz, so f becomes 11.5MHz. At this frequency there must be no slew-rate clipping, even at maximum amplitude. The bandwidth of the amplifier is also to be large enough to enable an accurate read-out of the first harmonics of the THD test.

The THD is to be below -60dB - 20log(11.8/17.5) = -57dB = 0.15%. The second term relates the harmonic power to v_{inmax} instead of A. However, if the signal consists of an entire radio-band and the distortion power is not evenly distributed over the band, a lower THD-figure will be required.

6.3.7 Numerical Experiments

It is very difficult to measure the intermodulation distortion power, as the distortion and signal components are very hard, if not impossible, to separate from each other. This method was therefore validated by numerical experiments instead of real measurements. The experiments were carried out on a computer using the program MATLAB [9].

6.3.7.1 Static Nonlinearity Only

Two different nonlinearities with negative feedback were studied, a square-law (ideal MOSFET) and an exponential (ideal BJT). The harmonic distortion of these nonlinearities with negative feedback is treated in [10,11]. The clipping limits were set where the fundamental peak current was 75% of the quiescent current. One reason for not selecting 100% is that the exponential characteristic never reaches zero. Another reason is that if the limit is set too high, the high order terms might become dominant for large signals, and A might have to be adjusted. The results are shown in table 6.2. It can be seen that for both nonlinearities the THD-test gives a good estimate of the intermodulation.

Table 6.2 : IM and THD related to maximum fundamental signal before clipping, $\sigma=1/4$ and $A=2.5\sigma$.

type	A_{loop}	IM	THD
quadratic	10	0.26%	0.32%
(MOS)	100	0.031%	0.039%
exponential	24	0.26%	0.33%
(BJT)	210	0.031%	0.039%

6.3.7.2 Both Dynamic and Static Nonlinearity

A dominant pole amplifier with feedback was studied. Both the input and the output stages were nonlinear, see figure 6.8.

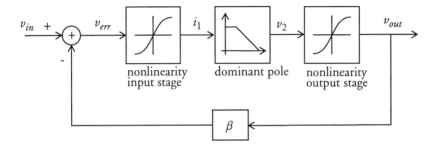

Figure 6.8: Amplifier model for the numerical experiment

The parameters were selected to resemble an audio power amplifier:

- Dominant pole frequency = 1kHz
- Input stage gain = 20mS
- Second stage gain = 200kΩ
- Output stage voltage gain = 1
- $\beta = 1/20$
- Maximum input stage output current, $i_1 = \pm 2$mA
- Maximum output voltage, $v_{out} = \pm 20$V
- Input stage nonlinearity before clipping: 3rd order compressive with 3rd order intercept point of 288mV referred to v_{err}
- Output stage nonlinearity before clipping: 3rd order compressive with 3rd order intercept point of 56V referred to v_2

From this set of parameters other parameters can be calculated:

- Bandwidth = 200kHz
- DC loop gain = 200
- Slew-rate = ± 2.5V/µs

The slew-rate is just large enough to produce an output signal of 20V at 20kHz. The result is that the slew-rate clipping distortion is smaller than the static clipping distortion for a wideband signal with f_w up to 20kHz * 1.73 = 34.6kHz.

To make the MATLAB-program simple, Forward Euler was used as an integration method. The input signal was generated with sufficiently small time steps to make the integration numerically stable.

A Gaussian signal with constant spectral density from DC to 20kHz and $\sigma=0.2$ was generated and sent through the nonlinear amplifier model. The signal was also sent through an amplifier model that was identical in every respect except that the nonlinearities were removed. The power of the difference between the outputs, which is equal to the power of the intermodulation distortion caused by the nonlinearities, was then calculated.

In the THD-test a sinusoid with the amplitude $2.8\sigma=0.56V$ and the frequency 20kHz/1.73=11.56kHz was generated and fed to the amplifier model. The THD-figure was found using a Fast Fourier Transform (FFT) on the output signal.

The distortion related to maximum amplitude was 0.041% for the Gaussian signal, and 0.074% for the sinusoid. The estimation is in this case pessimistic, but less than a factor 2 too large. The importance of correct frequency can be demonstrated by making THD-tests at the same amplitude but different frequencies. At 20kHz the THD was 0.099%, and at 2kHz it was 0.014%, indicating that the test frequency is approximately correct.

6.4 References

[1] H. Sjöland and S. Mattisson, 'Intermodulation Noise Related to THD in Wide-Band Amplifiers', *IEEE Trans. on Circuits and Systems, Part I*, pp. 180-183, Vol. 44, No. 2, Feb. 1997

[2] H. Sjöland and S. Mattisson, 'Intermodulation Noise Related to THD in Dynamic Nonlinear Wideband Amplifiers', *IEEE Trans. on Circuits and Systems, Part II*, pp. 873-875, Vol. 45, No. 7, July 1998

[3] A. Kristensson, *Design and Analysis of Integrated OTA-C Low-Pass Filters*. Thesis, LUTEDX/(TETE-7081)/1-108(1997)

[4] C. C. Enz, F. Krummenacher and E. A. Vittoz, 'An Analytical MOS Transistor Model Valid in All Regions of Operation and Dedicated to Low-Voltage and Low-Current Applications', *Analog Integrated Circuits and Signal Processing*, pp. 83-114, Vol. 8, No.1, July 1995

[5] R. A. Brockbank and C. A. A. Wass, 'Non-Linear Distortion in Transmission Systems', *J. Inst. Elec. Engrs.*, part III, pp. 45-56, March 1945

[6] P. Z. Peebles, *Probability, Random Variables, and Random Signal Principles, third edition*, McGraw-Hill, Singapore, International Editions 1993

[7] B. Westergren and L. Råde, *BETA Mathematics Handbook*. Lund: Studentlitteratur, 1993

[8] E. M. Cherry, 'Transient Intermodulation Distortion - Part I: Hard Nonlinearity', *IEEE Trans. on Acoustics, Speech, and Signal Processing*, vol. ASSP-29, pp. 137-146, Apr. 1981

[9] The MathWorks, Inc., *MATLAB Reference Guide*. Natick: 1994

[10] P. J. Baxandall, 'Audio power amplifier design – 5', *Wireless World*, vol. 84, pp. 53-56, Dec. 1978

[11] ____, 'Audio power amplifier design – 6', *Wireless World*, vol. 85, pp. 69-73, Feb. 1979

Chapter 7

Audio Power Amplifiers

This chapter deals with low voltage integrated audio power amplifiers. Such an amplifier can be used in portable equipment with modest output power requirements, such as mobile phones, portable computers or small radio receivers.

In the first section the different requirements of audio power amplifiers in general, and battery-powered integrated ones in particular, are described. This is followed by two sections dealing with the different alternatives that must be considered when choosing class of operation and semiconductor technology. Finally, there is a section describing two novel realisations of CMOS class AB amplifiers. This section is complete with measurement results and chip photos.

7.1 Requirements of Audio Power Amplifiers

The requirements of an audio power amplifier are, of course, different for different applications. There is for instance a huge difference in the requirements of a power amplifier intended for use in stereo equipment and one for a mobile phone. Even though the requirements are so different, an attempt is made to describe what is important for an audio power amplifier in general.

7.1.1 General Requirements

The power amplifier should not reduce the bandwidth of the system, where the ears should be included as a part of the system. Humans do not hear sounds below 20Hz or above 20kHz, so not much more than that is required of an audio amplifier. In for instance an amplifier in a telephone or an amplifier used for just bass frequencies (such as in an active subwoofer), a lower bandwidth is sufficient.

An audio power amplifier is often required to be able to drive different loudspeakers. The impedance of one loudspeaker can be rather different from another, and an amplifier must be stable against self-oscillation for all loads that it can be subjected to.

To achieve a high sound quality, the distortion must be low over the entire operating bandwidth. How low depends on the application, where the hardest requirements are found in stereo equipment. The ear is differently sensitive for different types of distortion, and it is very hard to give any exact figures of the amount of distortion that can be accepted.

The output power requirement is the parameter that probably varies the most between different applications. The difference between, for instance, an amplifier used at a rock concert and one in a walkman is several orders of magnitude.

To get a nice frequency response most loudspeakers require the amplifier to have a low output impedance (voltage output). If the impedance is too high there will be a hump at low frequencies.

Another important parameter is efficiency. In some high-cost high-performance amplifiers the efficiency can be very low in order to achieve, for instance, low distortion. In very high output power amplifiers, however, the efficiency must be high in order to reduce the size and the cost of cooling. In battery-powered equipment the efficiency is important to achieve a long battery life.

7.1.2 Special Requirements of Integrated Battery-Powered Audio Power Amplifiers

As already mentioned among the general requirements, high efficiency is critical for a battery-powered amplifier. This will affect the choice of class of operation, see next section.

The number of external components should be kept at a minimum. It is best if the entire amplifier can be fully integrated; that is, the phase-compensation should be made without the use of inductors or large capacitors that can not be integrated.

The supply voltage requirements should be flexible, so an ability to operate at very low supply voltages is desirable. The output voltage swing and thereby the output power is, however, limited by the supply voltage. To avoid unnecessarily losing output power capability when the supply voltage is low, it is important that the maximum output voltage swing be as large as possible.

7.2 Class of Operation

When designing an amplifier the class of operation is a key decision. The different classes of operation are described in section 5.1. Since an audio amplifier is a wideband amplifier, classes C, D, E and F can be excluded from the discussion immediately.

In a battery-powered amplifier the power consumption is critical, so class A operation can also be excluded. The remaining classes are then B, AB, curved AB, curved A and S, where the first four are implemented similarly.

7.2.1 Class AB

The power efficiency of an ideal class B amplifier is in section 5.1 shown to be 78.5% for a sinusoid with maximum amplitude before clipping. This is the maximum value, as the efficiency drops with amplitude.

An advantage of class (A)B and curved (A)B is that the implementations can be simple. It is also possible to achieve full integration (no external components), and the disturbances on other circuits can be low since switching is not used. The design examples shown later in this chapter demonstrate that it is possible to achieve low distortion, low supply voltage, good stability for different loads, complete integration and reasonable efficiency using (curved) class AB in standard CMOS.

7.2.2 Class S

If an efficiency higher than that of class B is required, switching must be used. The theoretical efficiency can in class S reach 100%. As semiconductor process development results in faster electronic devices each year, high speed switching becomes a more and more interesting alternative for audio power amplifiers in general, and battery-powered in particular. A simplified schematic of a class S output stage is shown in figure 7.1.

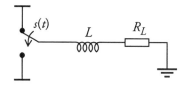

Figure 7.1: A simplified schematic of a class S output stage

The signal $s(t)$ controls the switch in such a way that the average voltage after the switch is equal to the desired output voltage. This is called pulse-width modulation [1].

The waveform after the switch contains the wanted audio frequency signal, but it also contains large high frequency components due to the switching. The inductor L forms a low pass filter together with the load resistance. This filter suppresses the high frequency components, ideally leaving just the audio signal.

Since ideally no energy is lost in the inductor or in the switches, all energy will be dissipated in the load, resulting in 100% efficiency. Unfortunately, energy is lost in practical implementations, particularly in the switches. Also the generation of the modulating waveform $s(t)$ and the driving of the switches require energy.

A major drawback of this type of amplifier is disturbances on other circuits. The high speed switching generates disturbances over a wide frequency spectrum. This can be particularly problematic if the amplifier is part of a system on a chip. Another disadvantage is the inductor, which due to its size can not be integrated on the chip. Perhaps it could be possible to utilize the inductor inherently present in an electrodynamic loudspeaker element. If this is tested, it is important that the disturbance problem is taken into account.

No amplifier designs using this technique are presented in this book, but interesting experiments can undoubtedly be made in this area.

7.3   CMOS vs. Bipolar

When designing an integrated amplifier one must decide which semiconductor technology (process) to use. The most commonly available processes today are CMOS, bipolar and BiCMOS. BiCMOS is a combination of bipolar and CMOS, combining their advantages. The drawback is that it requires more fabrication steps, and therefore becomes more expensive [2].

The CMOS and bipolar devices have different advantages and drawbacks. For audio power amplifiers the bipolar devices have a large advantage over CMOS in that they have higher transconductance for the same chip area and bias current. This is important, as audio power amplifiers must be able to drive low-impedance loads and still have a voltage gain of the output devices that is larger than unity. Also, in input stages the bipolar devices have advantages compared to CMOS. They have lower 1/f-noise and also lower offset voltage when used in a differential pair. Furthermore, it is possible to use a supply voltage as low as 1V with bipolar technology today [3].

An advantage of CMOS devices is that at audio frequencies they have a high input impedance, and thereby also current gain. It is also possible to build zero offset analog switches in CMOS, if for instance an input signal selector is desired on the same chip. Most important, however, is that the majority of the digital circuits are made in CMOS. If the analog parts are also made in CMOS, it becomes possible to create a system on a chip. If sufficient performance can be achieved in CMOS, it is preferred over a bipolar solution.

7.4 Two CMOS Class AB Designs

Two different class AB audio power amplifier designs have been built. Both are able to drive an 8Ω loudspeaker directly from the chip. The first design uses a Blomley class AB topology and is implemented in 1.2μm CMOS [4]. It is the first time this topology is used in an integrated CMOS power amplifier. The other design features a novel topology optimized for low voltage and can be operated with supply voltages as low as 1.5V [5]. It is implemented in a 0.8μm CMOS process.

Both the amplifiers are fully integrated, in that they use no external components for phase-compensation. No reference circuits are implemented, however, so the biasing relies on external references. This is, however, something that can easily be added to the circuits, making them truly fully integrated.

7.4.1 Output Stages

Both the amplifiers feature class AB output stages. The nonlinearity associated with the class AB operation could also have been placed in the input stage. This would have resulted in a large slew-rate, but created problems with offset voltage instead. Achieving a sufficiently large slew-rate is, however, easy also without a class AB input stage, so this solution was avoided due to the offset problems.

At low supply voltages it is important to maximize the output voltage swing. If the output devices are connected as source-followers, the swing is reduced by a threshold voltage per side compared to a common-source connection. Both the amplifiers therefore use common-source connected output transistors.

The output devices get very large in an amplifier designed for an 8Ω load. One reason is that the transconductance must be large enough to provide voltage gain when loaded with 8Ω. An even harder requirement is that the effective gate-source voltage must be kept low in order for the output devices to stay in saturation at high output voltages. This is very important if low distortion combined with high output voltage swing is to be achieved. If the length of the output devices is about 1μm, one should not be surprised if the required width of the P-device is more than 10mm. To build these very large devices, finger layout is employed. More about layout issues can be found in chapter 10.

7.4.1.1 Output Stage with Blomley Topology

The Blomley topology is based on the idea of splitting the signal into one positive and one negative half using a signal splitter. The different halves can then be amplified by separate class A amplifiers whose outputs finally are summed [6]. The class B nonlinearity is in this way placed at the signal splitter before the final stages. The concept is illustrated in figure 7.2.

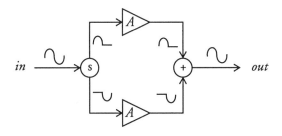

Figure 7.2: The Blomley topology

In this design current was chosen to represent the signals. Thereby the implementation of the signal splitter and the addition of the signals at the output become simple. An audio power amplifier must have a voltage output, however, but that is easily achieved by global feedback.

A property of current that is useful when building both signal splitters and adders, is that what goes in must come out. Assume the current splitter to have three terminals. The input current enters one of them. The sum of currents of the two other terminals is then identical to the input current. If these two currents are amplified in identical linear current amplifiers whose outputs are summed, no errors are introduced. The summing of the outputs is accomplished by simply connecting them together.

7.4 Two CMOS Class AB Designs

In figure 7.3 it is shown how the output stage is implemented using MOS transistors.

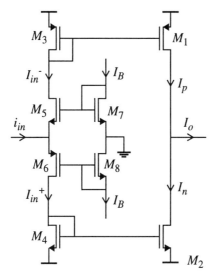

Figure 7.3: Schematic of the Blomley output stage

The signal splitter consists of M_5 and M_6. The current amplifiers are implemented as current mirrors (M_3, M_1 and M_4, M_2). Transistors M_7 and M_8 are used for biasing. The different functions are implemented using a surprisingly low number of devices. As mentioned before, the current entering a three terminal signal splitter must also come out. The current coming out is amplified in current mirrors, which are, ideally, linear. There is, therefore, potential for an output stage with this topology to be highly linear. A drawback is, however, that the output stage does not work properly for supply voltages less than $2(V_{GS}+V_{DS})$, which is twice the ultimate low voltage limit. A topology that is better in that respect is presented in the next section.

Ideally, the output stage is completely linear, but in practice a number of phenomena prevent this. In the practical signal splitter there is a fourth terminal, signal ground, to which signal current is lost through parasitic capacitances. Another problem is to get the correct input current to the splitter, since it has a nonlinear input impedance. The impedance of the stage preceding the signal splitter must therefore be very large to minimize the distortion. It is also important that the current mirrors have identical gain and bandwidth, so that both signal halves are treated equally. In addition, capacitances cause the current mirrors to become nonlinear at operating frequencies. The current mirrors also lose their linearity when the output devices leave saturation at large output voltages.

Nothing has so far been mentioned about how the signal splitter divides the signal current. This determines the appearance of the tail current diagram, see figure 5.2. The quadratic characteristic of the MOS devices will result in a curved class AB operation. The sum of the gate-source voltages of M_5 and M_6 is constant, resulting in the following equation for the currents in the curved region:

$$\sqrt{I_n} + \sqrt{I_p} = 2\sqrt{I_{BO}} \qquad I_n, I_p > 0 \qquad (7.1)$$

The tail currents can, after some algebra, be expressed in the output current I_o and the output bias current I_{BO}:

$$\begin{cases} I_p = \left(I_{BO} + \dfrac{I_o}{4}\right)^2 / I_{BO} \\ I_n = \left(I_{BO} - \dfrac{I_o}{4}\right)^2 / I_{BO} \end{cases} \quad |I_o| < 4I_b \qquad (7.2)$$

In the curved region the tail currents have a smooth quadratic form which will result in a low cross-over distortion. The stage will operate in the curved region for output currents up to four times the quiescent current. At larger currents one of the output stage halves turns off; that is, class B is entered.

7.4.1.2 A Low Voltage Output Stage

To approach the fundamental low voltage limit, another output stage topology was tested. Current mirrors are also used here at the output. Instead of the current-splitter they are, however, driven by common source stages, see figure 7.4.

No more than one gate-source voltage appears in series, which is a requirement for low voltage. If the current mirrors are linear with the same current gain and the transistors M_1 and M_2 are ideal square-law and perfectly complementary, the output stage is linear from V_{in} to I_{out} when both M_1 and M_2 conduct. This is based on the principle that the difference between two square-law functions is linear [7]. However, practical devices will not be perfectly complementary or square-law, resulting in nonlinearity. Another problem mentioned in the previous section is that the current mirrors can not be completely linear at operating frequencies.

7.4 Two CMOS Class AB Designs

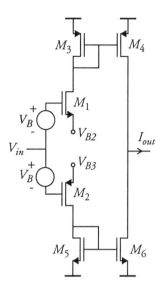

Figure 7.4: The basic topology of the low voltage output stage

At a certain input voltage, M_1 or M_2 will stop conducting, and the stage will enter class B. The gain of the M_1,M_2-stage will then increase as the input voltage rises, counteracting the gain reduction caused by the current mirrors at large output voltages. Just as for the Blomley stage, class B will be entered when the output current reaches four times the quiescent current. It is thus possible to bias the stage for (curved) class AB operation. The question that now arises is how to establish the bias voltages so that the output bias current is accurately defined. In figure 7.5 two different alternatives are shown.

The (a) alternative is the more accurate, but also more complicated. The operational amplifiers must provide a low output impedance. In fact, their entire purpose is to act as buffers with high input impedance and low output impedance. The output bias current becomes:

$$I_{BO} = I_B \cdot \frac{(W/L)_1}{(W/L)_7} \cdot \frac{(W/L)_4}{(W/L)_3} = I_B \cdot \frac{(W/L)_2}{(W/L)_8} \cdot \frac{(W/L)_6}{(W/L)_5} \quad (7.3)$$

The bias thus relies on matching of transistors, which can be very accurate.

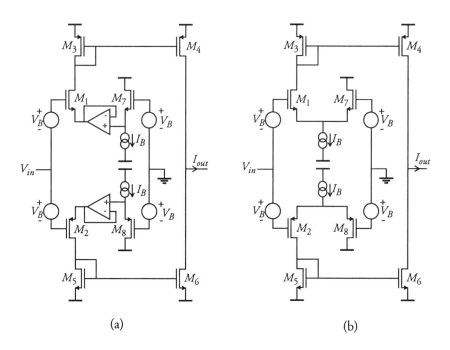

Figure 7.5: Different methods to establish V_{B2} and V_{B3}

In the (b) alternative, source followers are used to provide the low-impedance bias voltages to the sources of M_1 and M_2. The source follower transistors M_7 and M_8 must be much wider than M_1 and M_2 to provide a sufficiently low impedance. Another way to view this is that the class B nonlinearity is based on complementary mismatched differential pairs. The output bias current also here relies on matching, see equation (5.8).

The advantage of (b) is its simplicity. The large-signal behaviour, however, differs from that of (a) in that the maximum output current is limited to I_B multiplied by the gain of the current mirrors. This will result in increased distortion at large output currents, but perhaps it can be utilized as a short-circuit protection. In order to avoid unnecessary distortion, the maximum current is to be selected much larger than the largest current that will occur under normal operation.

The (b) alternative was selected for implementation because of its simplicity. The level-shifts were implemented as resistances through which DC-current was drawn.

7.4.2 Phase-Compensation

The phase-compensation of a fully integrated audio power amplifier is a highly demanding task. It is not possible to build a load stabilizing network to ensure a well-known resistive load at high frequencies, since such a network requires large coils and capacitors that can not be integrated. Therefore, the amplifier must be stable for loads with different resistive and capacitive parts. The resistance is allowed to be as small as 8Ω, and the capacitance as large as 100nF. This must be handled with good phase margin; that is, only modest overshoot is allowed on a step-response. Furthermore, to achieve high linearity the phase-compensation must not reduce the feedback around the output stage too much, and the amplifier stages must be loaded as little as possible by the phase-compensation.

The two amplifiers have different output stage topologies, and therefore different phase-compensation strategies are employed.

7.4.2.1 Phase-Compensation of the Blomley Amplifier

There are three significant poles in the output stage. One is due to the load capacitance, one occurs in the current mirrors and one in the current splitter. Since the output stage has a current output, the pole associated with the load can be located almost anywhere, and the open-loop gain magnitude is determined almost completely by the load.

The strategy is to employ local negative feedback around the output stage to fix its gain magnitude (transresistance) and to create a pole at a fixed frequency. The other poles are moved to higher frequencies by the local feedback. The fixed pole, which is created by a phantom-zero in the local feedback, is cancelled by a zero in the input stage, which also has a dominant pole, see figure 7.6.

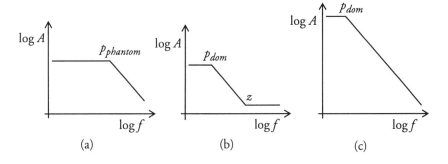

Figure 7.6: Simplified frequency responses of (a) the output stage (b) the input stage and (c) the combination of both

The preferred input signal of the output stage is a current and the desired output signal is a voltage, so the local feedback must be a transconductance. Another complication is that only small capacitors (with high impedance at low frequencies) can be integrated on the chip because of area limitations. An active feedback network is therefore used, figure 7.7.

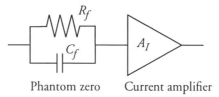

Figure 7.7: The local feedback network of the output stage

The current amplifier provides current gain and a current output (high impedance). The current gain enables the feedback network to work despite a small capacitance. This current gain multiplied by that of the output stage, A_{Itot}, must not be smaller than the ratio of the maximum load capacitance to the capacitance of the feedback network, equation (7.4). If the current gain is smaller, the capacitive part of the feedback will not have any significant effect under maximum capacitive load conditions.

$$A_{Itot} > \frac{C_{Lmax}}{C_f} \tag{7.4}$$

The amplifier must be able to operate with load capacitances up to 100nF. With a capacitance of 6pF in the feedback network, equation (7.4) gives that the total current gain must be at least 17,000. Most of the gain should be provided by the output stage, as this reduces its input current and thereby also the quiescent current of the class A amplifiers of the active feedback and the input stage. The gain required from the active feedback is then also reduced, making it easier to achieve high enough frequencies for its poles not to affect the loop stability.

There are, however, not just advantages in making the current gain of the output stage large. Above the frequency of the current mirror poles the amplifier will be less efficient, since the current mirrors will smooth out the signals from the splitter by filtering out high harmonics. If just the DC-component and the fundamental would pass, the amplifier would work in push-pull class A, with an accordingly low efficiency. Despite the higher current, there will be a lot of distortion for signals with frequencies above the pole because of bad matching and nonlinearities of the current mirrors. A compromise thus has to be made when choosing the gain of the output stage.

7.4 Two CMOS Class AB Designs

The loop gain (local feedback) of the output stage for different loads is shown in figure 7.8. To avoid instability in the worst case (c), additional zeros are inserted in the loop by shunting some devices with capacitances.

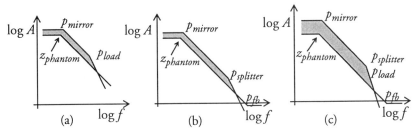

Figure 7.8: The demanded gain and the open-loop gain of the output stage with different loads. The shaded area represents the loop gain.
(a) Resistive and capacitive load (b) pure resistive load
(c) pure small capacitive load, worst case

7.4.2.2 Phase-Compensation of the Low Voltage Amplifier

For the low voltage output stage, the reversed nested Miller [8] compensation strategy was chosen. In figure 7.9 the output stage is shown together with the phase-compensation capacitors c_{m1} and c_{m2}.

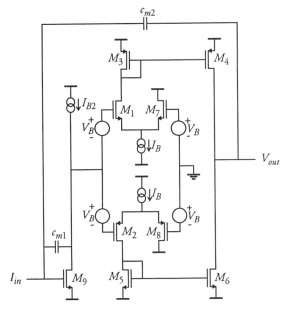

Figure 7.9: The output stage with reversed nested Miller compensation

The inner Miller loop encloses just transistor M_9, and the outer loop encloses both M_9 and the output stage. A drawback is that the output stage is not linearized by the inner loop. The nested Miller compensation does not have this disadvantage, but it results in a smaller bandwidth when large load capacitances are allowed [8].

The stability and linearity are essentially determined by the feedback around the output stage. In figure 7.10 the asymptotic voltage gain of the output stage together with the inverse of the feedback factor β is shown in the same diagram.

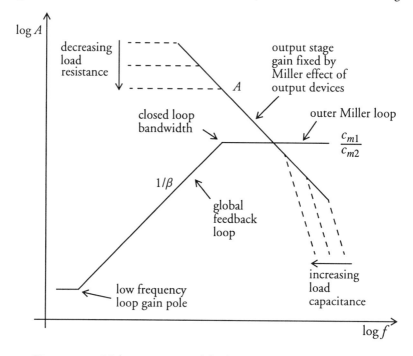

Figure 7.10: Voltage gain A and feedback factor β of the output stage

As can be seen in figure 7.10, the Miller effect of the output devices helps us to fix the gain of the output stage. To take full advantage of this, the current mirrors must have large current gain, just as they had in the Blomley amplifier. Due to the limited size of capacitors, it is not possible to increase the Miller effect substantially by shunting the very large output devices with capacitors.

As shown in [8] the high frequency β of the output stage is c_{m2}/c_{m1}. By selecting this ratio it is possible to get an appropriate amount of high frequency feedback. The closed-loop bandwidth of the amplifier is determined by c_{m2} and the transconductance of the input stage. Since the noise performance puts a lower limit

7.4 Two CMOS Class AB Designs

on the input stage transconductance, minimum values of the capacitances can then be calculated. As the capacitors occupy chip area and load the amplifier stages, the minimum values are used. The capacitors and the bias current of the input stage determine the slew-rate. This does not necessarily result in a design conflict, as different dimensions of the input devices enable the bias current to be selected differently without changing the transconductance.

There is an interesting similarity between figure 7.10 and the loop gain around the output stage of the perhaps most common audio power amplifier topology. This is a global feedback topology, using a differential input stage followed by a Miller compensated CE-stage, which drives a local feedback output stage with unity voltage gain [9]. As in our topology, the output stage then has a constant β at high frequencies. Due to the Miller integrator present in both topologies, the global feedback decreases at single-pole rate at frequencies below the closed-loop bandwidth.

There is also a similarity between the phase-compensation of this amplifier and the Blomley amplifier, in that both rely on local feedback around the output stage, combined with a high-gain stage with a dominant pole preceding the output stage.

7.4.3 Input Stages

To enable subtraction of the global feedback signal from the input signal, the input stages have both inverting and non-inverting inputs. In both topologies the required transfer function is a transconductance. The input stages are, however, quite different, since the Blomley amplifier requires a certain frequency characteristic and the low voltage input stage uses level-shifts.

7.4.3.1 Blomley Amplifier Input Stage

To get the frequency response with a left halfplane zero of figure 7.6b, a parallel connection of two transconductance amplifiers is used as input stage, figure 7.11. The addition of the outputs is simple, since the signal is a current.

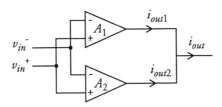

Figure 7.11: The input stage of the Blomley amplifier

The frequency response of the amplifiers is shown in figure 7.12. The solid line is the frequency response of the combination.

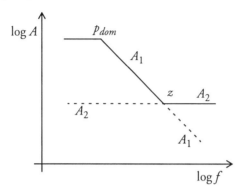

Figure 7.12: The frequency response of the input stage.

Amplifier A_1 has a large low-frequency gain and a dominant pole at a low frequency. This can be accomplished by an uncompensated two-stage operation-amplifier, which has a high impedance node that can be loaded by a capacitor to create the dominant pole. Amplifier A_2 has low gain and must not have any low-frequency poles, which can be accomplished by a one-stage operation-amplifier.

7.4.3.2 Low Voltage Input Stage

The input stage of the low voltage amplifier is quite simple. It just consists of a differential pair with a current-mirror load, figure 7.13.

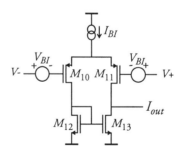

Figure 7.13: The low voltage input stage

7.4 Two CMOS Class AB Designs

In order to enable operation on a ±0.75V supply with the input referred to ground, level-shifting at the inputs is needed to make the input devices conduct. As in the output stage, the level-shifting is accomplished using resistors through which DC-current is drawn. Since the noise usually is not critical in a power amplifier, it is acceptable to use resistors at the inputs. If the voltage gain is to be larger than unity, resistors are anyway used in the feedback network connected to one of the inputs. However, to avoid excessive noise, the resistance must be kept down. Good matching is critical, as a mismatch of the level-shifts will appear as an offset voltage at the input. More about matching issues can be found in chapter 10.

7.4.4 Simulation and Measurement Results

To estimate the performance of the amplifiers before they were sent to fabrication, the circuits were simulated. Particularly the linearity and the stability with different loads were examined. The transistors were modelled using the MOS2-model.

When the circuits were fabricated, measurements were made. The distortion and noise were measured using an audio analyser instrument. The other measurements were made using multimeters, a signal generator and an oscilloscope.

7.4.4.1 The Blomley Amplifier Results

The total schematic that was used in the final simulations is shown in figure 7.14. It is an operational amplifier designed to be used with the gain $A_v = 5$ and the supply voltage 5V.

The current gain of the output current mirrors is about 2500. Simulations were made to find their bandwidth, which was about 200kHz, depending on the operating point. The N-channel output transistor was made longer than minimum length to match the bandwidth of the two output current mirrors. This both reduces the distortion and makes the phase-compensation easier.

Calculations combined with simulations were used to place the dominant pole at 2kHz. The phantom-zero and the zero of the input stage were placed at 100kHz. In both simulations and measurements the amplifier was stable with load resistances from 8Ω to infinity and capacitances from zero to 100nF. The voltage gain was equal to 5. In fact, measurements were made with capacitances up to 1μF, and the amplifier was also stable with that load, but the step response showed a lot of ringing.

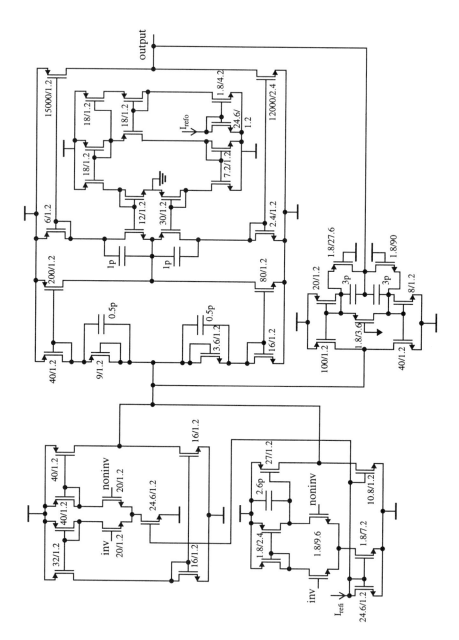

Figure 7.14: Total schematic of the power amplifier with Blomley topology

7.4 Two CMOS Class AB Designs

Some results:

- maximum simulated output swing with 8Ω load = $4.4V_{pp}$. The measured swing was slightly less at $4.2V_{pp}$.
- measured total quiescent power = 60mW (12mA)
- measured bandwidth with 8Ω, 100nF load = 260kHz
- measured output noise in the 0-30kHz band ≈ $0.25mV_{rms}$

The distortion was simulated and measured:

Table 7.1 : Simulated (S) and measured (M) THD vs. output amplitude and frequency, $R_L=8\Omega$, $A_v=5$.

	1kHz	10kHz	20kHz
100mV$_{pp}$	S: 0.0077%	S: 0.051%	S: 0.10%
1V$_{pp}$	S: 0.011%	S: 0.096%	S: 0.28%
	M: ——	M: 0.3%	M: 0.5%
3V$_{pp}$	S: 0.028%	S: 0.24%	S: 0.53%
	M: 0.08%	M: 0.5%	M: 0.8%
4V$_{pp}$	S: 0.12%	S: 0.87%	S: 1.4%
	M: 0.2%	M: 1.3%	M: 1.6%

The high distortion at large output voltages is caused by the output transistors leaving saturation, and thereby causing the output current mirrors to lose their high linearity. Other fully integrated CMOS low voltage amplifiers designed for 8Ω loudspeaker loads are difficult to find, so it is hard to compare the distortion figures to other designs. The low voltage amplifier of this section, however, has better linearity when the same supply voltage (5V) is used. It is built in a better process (0.8 instead of 1.2μm), but is optimized for very low supply-voltages. To make the comparison more fair it must be mentioned that the Blomley amplifier uses a slightly smaller quiescent current.

7.4.4.2 The Low Voltage Amplifier Results

The total schematic of the low voltage amplifier is shown in figure 7.15. This is what was simulated, sent to fabrication and measured.

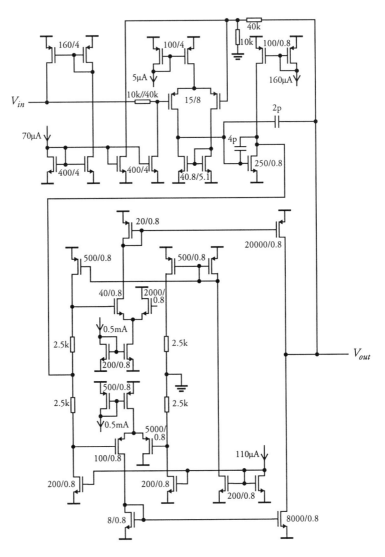

Figure 7.15: Total schematic of the low voltage audio power amplifier

7.4 Two CMOS Class AB Designs

The measurements and simulations were made with 1.5V, 3V and 5V supply voltage. The reference currents were set as indicated in the schematic. The total quiescent current in the simulations was 8.7mA at 1.5V supply and 13.7mA at 5V supply. In the measurements the current was slightly larger, 10.4mA and 22mA.

The voltage gain was set to 5 by an internal resistive feedback network. In both simulations and measurements, the voltage gain was very close to 5. The amplifier was stable for different supply voltages and load impedances. Load capacitances up to 1µF were also tested in the measurements for this amplifier. The result was similar to that of the Blomley amplifier; that is, the phase-compensation worked very well.

The clipping limit, defined as where the THD is 1% at 1kHz with 8Ω load, was measured. It was $4.3V_{pp}$ at 5V supply, $2.4V_{pp}$ at 3V and $1.0V_{pp}$ at 1.5V. In the simulations, the noise was $100nV/(Hz)^{1/2}$ referred to the input. The measured noise was slightly higher at $130nV/(Hz)^{1/2}$. Simulations were made to find the distortion of the amplifier. The results are presented in the graphs of figure 7.16 and figure 7.17.

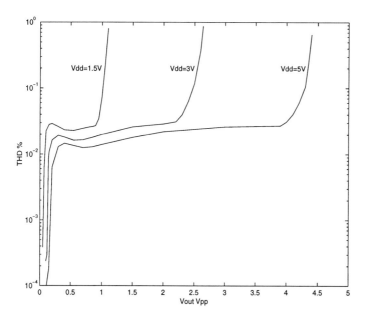

Figure 7.16: THD at 1kHz 8Ω vs output amplitude and supply voltage

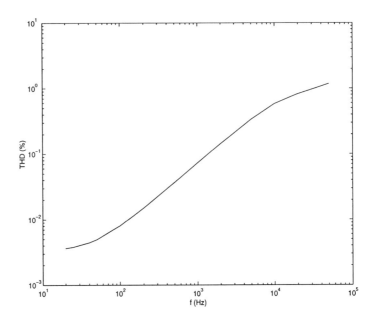

Figure 7.17: THD vs. frequency at $1V_{pp}$ in 8Ω and 1.5V supply

At low output voltages there is a hump in the distortion curves of figure 7.16. This is caused by a discontinuity in the transconductance of the transistor model (MOS2). This discontinuity occurs at the transition between weak and strong inversion, and results in additional distortion when the amplifier enters class B, as M_1 or M_2 then leaves strong inversion. The real distortion should therefore be lower and without the hump.

The THD (+noise) measurements are presented in table 7.2. The noise is also shown to indicate the accuracy of the THD measurements. At 1kHz the THD measurements are dominated by the noise, except for amplitudes close to the clipping limit. The 10kHz measurements are more accurate (less disturbed by noise). To estimate the THD at 1kHz, the 10kHz figure can be divided by 10, as the loop gain is 10 times larger at 1kHz. This is supported by figure 7.17. Using this estimate, the measured distortion is comparable to the simulations in figure 7.16.

7.4 Two CMOS Class AB Designs

Table 7.2 : Measured THD+Noise and Noise at 8Ω load

V_{out}	V_{DD}	THD@1kHz	THD@10kHz	Noise
0.6V_{pp}	1.5V	0.13%	0.44%	0.093%
0.8V_{pp}	1.5V	0.17%	0.49%	0.070%
1.5V_{pp}	3V	0.047%	0.24%	0.037%
2V_{pp}	3V	0.038%	0.23%	0.028%
2V_{pp}	5V	0.033%	0.15%	0.028%
3V_{pp}	5V	0.025%	0.16%	0.019%
4V_{pp}	5V	0.076%	0.61%	0.014%

The amplifier performs well in both simulations and measurements. It shows good stability, low distortion and can be used with supply voltages down to 1.5V.

7.4.5 Layout and Chip Photos

In order to handle the peak currents of up to 300mA at 5V supply and 8Ω load, several pads are used in parallel for the output and the supply voltages. The required current capability also makes the output devices so large that they dominate the layout. The layout work was therefore concentrated on them.

Since standard digital processes are used, no extra layer of polysilicon is available for creating capacitors. The capacitors are therefore implemented using the two metal layers and the single polysilicon layer. This is not area-efficient, so care must be taken to minimize the chip area. At some places in the layouts large metal areas are anyway present to conduct current, so by using them as capacitors as well, chip area can be saved.

Chip photos of the two designs are shown in figure 7.18 and figure 7.19.

110   7 Audio Power Amplifiers

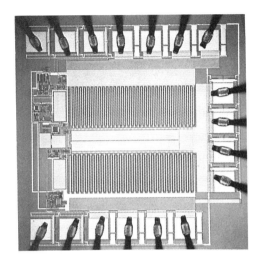

Figure 7.18: Micrograph of the 8Ω amplifier with Blomley topology

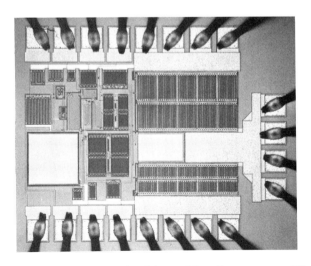

Figure 7.19: Micrograph of the 1.5V audio power amplifier

7.5 References

[1] P. T. Krein, *Elements of Power Electronics*, Oxford University Press, 1998

[2] M. Ismail and T. Fiez, *Analog VLSI – Signal and Information Processing*, McGraw-Hill, 1994

[3] M. J. Fonderie and J. H. Huijsing, *Deisgn of Low-Voltage Bipolar Operational Amplifiers*, Kluwer Academic Publishers, 1993

[4] H. Sjöland and S. Mattisson, 'A Novel Class AB CMOS Power Amplifier', *Analog Integrated Circuits and Signal Processing*, pp. 49-58, Jan. 1997

[5] H. Sjöland, 'A 1.5V Class AB CMOS Power Amplifer', Proceedings Nor-Chip '97, pp. 180-186, Tallinn, Estonia, Nov. 1997

[6] P. Blomley, 'New Approach to Class B Amplifier Design', *Wireless World*, pp. 57-61, Feb. 1971

[7] I. Hegglun, 'Square Law Rules in Audio Power.' *Electronics World + Wireless World*, Aug. 1993, pp. 630-634

[8] R. G. H. Eschauzier and J.H. Huijsing, *Frequency Compensation Techniques for Low-Power Operational Amplifiers*, Kluwer Academic Publishers, 1995

[9] D. Self, 'Distortion in Power Amplifiers, 1: The Sources of Distortion', *Electronics World + Wireless World*, Aug. 1993, pp. 630-634

Chapter 8

Wideband IF Amplifiers

Wideband IF amplifiers can be used in communication systems featuring high data rates and thereby wide channel-bandwidths. They can also be used to amplify a radio band consisting of several narrow-band channels. This was discussed in section 1.1, where it was also explained why the linearity is critical for a wideband IF amplifier.

This is a new application, for which we want to find a suitable topology. Such a topology is found and tested in this chapter. It is tested by building high-performance amplifiers in both CMOS and bipolar technology. The description of the amplifiers is complete with measurement results and chip photos. The amplifiers have also been presented in IEEE Journal of Solid-State Circuits [1,2] and I am grateful to IEEE for letting me use the material for this chapter.

8.1 Performance Requirements

We try to make amplifiers with high performance in terms of noise and linearity. If then the requirements turn out to be lower in a specific application, the power consumption can be decreased.

The operating frequency band is DC to 20MHz, which allows, for instance, 100 GSM channels to be amplified simultaneously.

If the intermodulation distortion is lower than the noise, there is no improvement in reducing the nonlinearity further. How much THD this corresponds to can be calculated using the statistical method described in chapter 6. First we must, however, decide which signal level to use and how much noise to accept.

The voltage gain is to be 100. The maximum output voltage swing is about $6V_{pp}$, which is achieved by using a differential output. The source impedance is two times 100Ω (differential input). With a noise figure of 3dB, the maximum signal to noise ratio at the output then becomes 65dB.

To get a clipping distortion that is below the noise floor, the variance (signal level) is to be selected low enough. A variance equal to the peak voltage divided by 4.5 results in a clipping distortion that is about 5dB below the noise. Since the odd nonlinearities will dominate (a differential topology will be used to cancel the even ones) the variance should be multiplied by 2.8 to find the amplitude of the THD-test. The output amplitude then becomes $6V_{pp} \cdot 2.8/4.5 = 3.75V_{pp}$. The frequency is to be 20MHz divided by the square root of 3, which is equal to 11.5MHz. The THD is to be below $-65-20\log(2.8/4.5)=-61dB=0.09\%$ for the intermodulation distortion to be below the noise floor. If we assume 100 channels and a distribution where all distortion falls into one channel, which is not realistic, the distortion requirement instead becomes 0.009%. A more reasonable estimate is that the THD is to be below about 0.03%. To conclude, the THD is to be below about 0.03% at $3.75V_{pp}$ output amplitude at 11.5MHz.

The impedance of the source has been specified but not that of the load. As the amplifier is intended for driving an on-chip load (AD-converter), it does not have to be able to drive loads with low impedance. The amplifiers are thus designed for a $1k\Omega$ load. They must also be able to handle a few pF capacitive load.

8.2 The Topology

One of the primary requirements of a wideband IF amplifier is high linearity. As discussed in chapter 3, this can be achieved in different ways. The operating frequencies are moderate, so it is possible to use negative feedback to linearize the amplifier. Negative feedback has the advantage that it does not require close matching or tuning to perform well.

The output stages are to be as linear as possible before negative feedback is applied. Push-pull class A operation was therefore chosen for the output stages. To achieve the highest linearity it is necessary to use a suitable phase-compensation technique. The nested Miller technique [3-5] was chosen, as it provides the maximum feedback around the output stage. The amplifier is made fully differential to cancel out even order nonlinearity.

Also the noise is important for this type of amplifier. The noise performance is almost completely determined by the input stage. Since the design is to be fully differential, a differential pair without any feedback is the best input stage concerning noise. In bipolar technology, however, some feedback must be accepted to increase the linearity of the input stage (input signal handling capability).

Since the amplifier is fully differential, a common-mode feedback is needed. The complete topology is shown in figure 8.1.

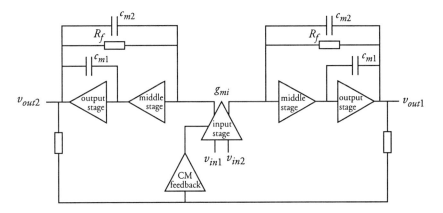

Figure 8.1: Topology of the wideband IF amplifiers

The figure shows a double-nested Miller topology. In order for the phase-compensation to work, the middle stages must be non-inverting and the output stages inverting. In addition, the input stage and the middle stages must not have too low an output impedance; that is, they must have current outputs. The overall (differential) voltage gain becomes equal to g_{mi} multiplied by R_f.

8.3 Output Stages

In the previous section it was decided to use push-pull class A output stages to achieve high linearity. It was also decided to use a nested Miller topology that requires the output stages to be inverting. The phase-compensation also requires the output stages to have no more than two significant poles. To reduce the number of poles, the output stages are to be kept as simple as possible.

In this section it is shown how these inverting push-pull class A output stages can be implemented both in CMOS and bipolar technology.

8.3.1 CMOS Output Stages

In CMOS both N and P devices are available. This enables a simple implementation of the inverting push-pull stage based on a CMOS inverter, see figure 8.2.

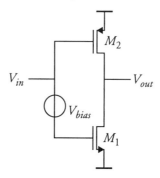

Figure 8.2: The CMOS output stage

This stage may look very simple, but conflicting requirements make the selection of dimensions and quiescent current of the output transistors complicated.

A certain minimum quiescent current is required to ensure class A operation at all output voltages (with specified load). The quiescent current is, however, limited by power consumption requirements. The quadratic characteristic of the MOS-devices can be exploited by using a quiescent current that is just one fourth of the peak output current, instead of one half as for conventional push-pull class A. The stage then operates in curved class A, and if the devices are matched and perfectly square-law, it is completely linear [6].

Large output voltage swing with low distortion requires low effective gate-source voltages, resulting in wide transistors. If the transistors are very wide, however, the driving stage is loaded too much. This can also be regarded as the current gain of the output stage getting too low.

The inner feedback loop is formed by the output stage and a Miller capacitor. The desired transfer of this feedback stage is current to voltage with a single dominant pole. The frequency of the second pole must be sufficiently high not to cause stability problems with the outer loop. This frequency is proportional to the transconductance of the output stage, which thereby also must be sufficiently large.

A reasonable compromise must made when choosing the bias point, taking all the above requirements into account. The bias is then established using the circuit of figure 8.3.

8.3 Output Stages

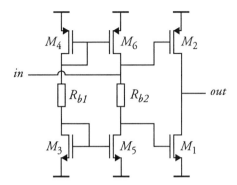

Figure 8.3: The CMOS output stage including bias circuit

Transistors M_3 to M_6 and the resistors implement the floating voltage source V_{bias} of figure 8.2. The current gain of the mirrors M_3,M_5 and M_4,M_6 is made equal to the ratio R_{b1}/R_{b2}. Thereby the voltage across the resistors becomes the same. All transistors have the same length and transistors M_1 and M_2 are made N times wider than M_3 and M_4, respectively. At zero output current, just the quiescent current flows through the output devices. Since the voltage across the resistors is the same, M_2 and M_4 then get the same gate-voltage, and so do M_1 and M_3. The quiescent current of the output devices thus becomes equal to that of M_3 and M_4 multiplied by N.

8.3.2 Bipolar Output Stage

In bipolar processes npn-devices often have superior performance to pnp. An output stage using just npn-devices was therefore designed. It is possible to design a push-pull stage without pnp-devices, but it is no longer sufficient with just two transistors. A drawback is that the output voltage swing is reduced by almost one volt, as one of the output devices must be connected as an emitter follower, see figure 8.4. The stage is identical to that of figure 5.10.

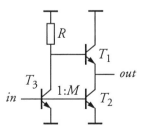

Figure 8.4: The bipolar output stage

As the stage will operate in class A, emitter-follower T_1 will conduct all the time so that the output voltage essentially will follow its base-voltage. The base-voltage is supplied by the inverting stage consisting of T_3 and R_1. The signal is thus inverted from input to output, just as the phase-compensation requires. Since transistor T_2 also is inverting, it provides the negative tail current in a phase appropriate for push-pull operation. The output quiescent current can be calculated using equation (5.10).

8.4 Input Stages

The topology chosen requires the input stage to be a differential transconductance stage with possibility to control the common-mode output current.

The input stage must be able to handle the input signal without contributing significantly to the total distortion; that is, its distortion must be less than that of the output stage. In addition, its noise must be low, since the noise performance of the entire amplifier is determined mainly by the noise of the input stage. Furthermore, the low noise and distortion must be achieved without excessive power consumption. More about noise issues in both bipolar and CMOS technology can be found in [7].

8.4.1 CMOS Input Stage

The CMOS input stage essentially consists of a differential pair. The common-mode control is achieved by loading the pair with controllable current sources, see figure 8.5.

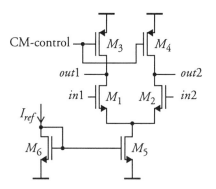

Figure 8.5: The CMOS input stage

In order to avoid offset problems, M_1 and M_2 must be closely matched, and so must also M_3 and M_4. In chapter 10 it is discussed how this can be achieved by

8.4 Input Stages

proper layout. The main concerns when choosing the bias current and device dimensions are noise, distortion and power consumption.

Let's start with the noise. We are concerned about differential noise, so the noise of M_5, M_6 and the noise at the CM-control input are not important as they enter the circuit in a common node. What is important instead is the noise of M_1 to M_4. To minimize the noise the input devices M_1 and M_2 are to have as high a transconductance as possible. This means that they must be very wide in order to utilize the current well. To minimize their noise contribution, transistors M_3 and M_4, on the other hand, are to have a transconductance much lower than that of the input devices. This is helped by the fact that M_3 and M_4 are P-devices. In addition, a lower width to length ratio is used for M_3 and M_4.

The next issue is linearity. The input stage must be able to handle the input signal without distorting more than the output stage. The voltage gain of the amplifier is 100, resulting in input signal amplitudes of up to about fifty millivolts (peak to peak), which can be handled by our simple differential pair. The input referred linearity increases with increased effective gate-source voltage. This contradicts the requirement for low noise combined with low current. As usual, a compromise must be made. In this case low noise and distortion were of higher priority than low power.

8.4.2 Bipolar Input Stage

The bipolar input stage also consists of a differential pair. In order to achieve the required linearity, however, degeneration (feedback) must be used. No pnp-devices are used, so the stage is resistor loaded. The common-mode control must thereby be located in the tail current sink. Two different input stages are shown in figure 8.6.

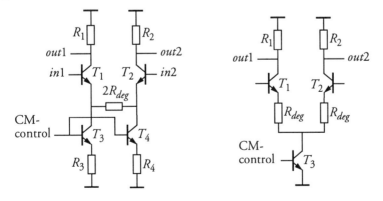

Figure 8.6: Two different bipolar input stages

Degeneration is needed since the input (voltage) referred nonlinearity of a bipolar transistor is independent of the bias level. The bias current can thus not be increased to allow larger input voltages as it can for MOS. With a peak input signal amplitude about v_T (the thermal voltage), degeneration is unavoidable.

The input stage to the right in figure 8.6 is the preferred one. In the left stage the noise of T_3 and T_4 do not enter the circuit in a common node, resulting in decreased noise performance. Since this was not realized before the circuit was sent for fabrication, however, the measured and simulated results are for the left stage. As for MOS, the devices of the input stage must be well matched. This applies to both transistors and resistors.

The linearity gets better with increased quiescent current and degeneration resistance. The noise performance, on the other hand, gets worse when the resistance is increased and has an optimum at a relatively large quiescent current. As for the MOS input stage a compromise must be found between current consumption, noise and distortion. Also in this case a relatively large current was used in order to achieve low noise and distortion.

8.5 Middle Stage

The purpose of the middle stage is to provide gain, which is used to increase the feedback around the output stage, thereby making it more linear. This stage must be non-inverting for the nested-Miller phase-compensation to work. It is similarly implemented in both the CMOS and bipolar circuits, figure 8.7.

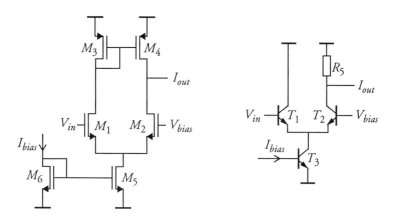

Figure 8.7: CMOS and bipolar middle stage

The CMOS stage has the advantage of the gain being doubled by a current mirror. Such a mirror can not be implemented in npn-only bipolar technology, because of the lack of pnp-devices. The bipolar devices have, however, higher transconductance than MOS for the same amount of bias current. The result is that the bipolar stage still has the highest gain for the same power consumption.

8.6 Common-Mode Feedback

Since the topology is fully differential, the common-mode output voltage must be controlled. This is done by a common-mode feedback circuit that senses the common-mode output voltage and controls it by adjusting the input stage.

The common-mode feedback circuits operate essentially in the same way in the CMOS and bipolar implementations. The common-mode output is sensed by two identical resistors connected in series between the two outputs. The node between the two resistors provides the common-mode voltage. This is then compared to the desired value, amplified and fed to the CM-control of the input stage. To make the feedback loop stable, a dominant pole is included in the circuit.

The CMOS implementation is shown in figure 8.8. The resistors sense the common-mode voltage, which is then compared to the reference value V_{CMref} by the differential pair M_1, M_2. Transistor M_8 is connected as a capacitance to create the dominant pole. By using a transistor as capacitor, a large capacitance per area can be achieved.

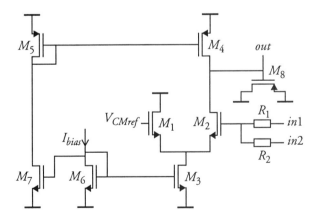

Figure 8.8: CMOS common-mode feedback circuit

The bipolar implementation is shown in figure 8.9. Resistors R_1 and R_2 sense the common-mode voltage. Resistor R_3 lowers its level so that it can be compared to the base-emitter voltage of T_1. The phase-compensation is performed by the Miller capacitor C_c aided by resistor R_4.

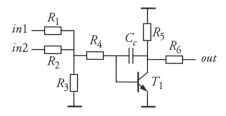

Figure 8.9: Bipolar common-mode feedback circuit

8.7 Simulations and Measurements

Both the CMOS and the bipolar circuit were simulated, sent to fabrication and measured. The CMOS circuit was fabricated in a standard 0.8μm CMOS process and the bipolar circuit used the bipolar part of a 0.8μm BiCMOS process. The results presented in this section show that the topology chosen is appropriate for both technologies. For even higher performance, a circuit could be made in BiCMOS, combining the advantages of the CMOS and bipolar technology. The only drawback would be increased cost.

All simulations were made using the simulator Spectre. The MOS-devices were simulated using the MOS2 model. The simulations included the layout parasitics and bondwire inductances.

The circuits were delivered in surface mount packages. Therefore, printed circuit boards (PCB) were made for the measurements. The circuits, bias components, decoupling capacitors and contacts (SMA) for the signals were mounted on the PCBs.

First the DC (bias) levels were measured using multimeters. Signals were then injected at the input. A signal generator followed by a signal splitter was used to create the differential input signal. An oscilloscope was used to measure the voltage gain at one frequency. A spectrum analyser was then connected the output, and the gain at other frequencies were measured relative to that at the first one. The frequency response could then be drawn. With the spectrum analyser connected to the output, the harmonics were studied and then the THD was calculated. A low-pass filter was inserted after the signal generator, filtering out its distortion, so that it should not spoil the measurement.

8.7 Simulations and Measurements

When the noise was to be measured, resistors were connected to the inputs. The noise at the outputs was measured using a spectrum analyser with a noise marker function. Some additional uncertainties occur when the noise is to be related back to the input by dividing by the voltage gain.

8.7.1 Schematics

The complete schematics with device parameters are shown in figures 8.10 and 8.11. This is what was simulated and sent to fabrication.

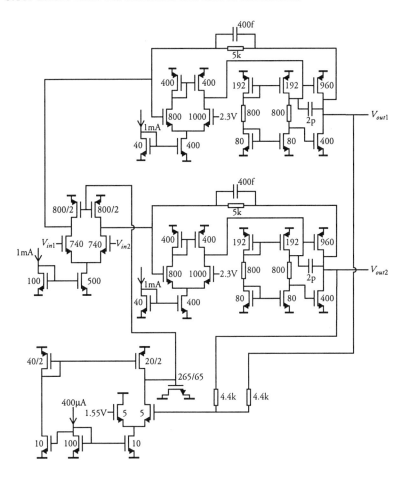

Figure 8.10: The entire CMOS schematic with device parameters. Unless specified, the transistor length is 0.8µm.

There is little to comment about the MOS schematic. In the middle stage, however, the input device is made slightly smaller than the other device in the differential pair. This is done to reduce the capacitance of the node between the input stage and the middle stage, as this capacitance reduces the loop gain and thereby increases the distortion.

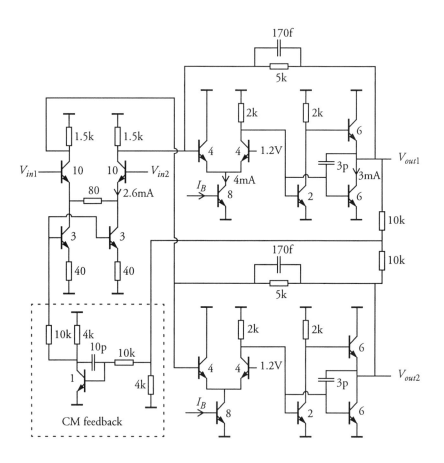

Figure 8.11: The entire bipolar schematic with parameters. The numbers at the transistors are the number of unit devices used. Each unit device has a 15µm x 1µm emitter.

The bipolar schematic is also straightforward. The input stage is, as can be seen, not the one with the lowest noise. If the chip were to be remanufactured, the input stage would therefore be changed.

8.7.2 Results

Some of the most important simulation and measurement results are summarized in table 8.1.

Table 8.1 : Some simulation and measurement results

	Simulated CMOS	Measured CMOS	Simulated Bipolar	Measured Bipolar
Quiescent power	154 mW	130 mW	133 mW	122 mW
Supply voltage	3.3 V	3.3 V	5 V	5 V
Voltage gain	100	100	96	100
Bandwidth	140 MHz	---	160 MHz	190 MHz
NF (R_s=2*100Ω) @ 10MHz	1.9 dB	4.0 dB	3.6 dB	5.4 dB
IP_3 @ 20MHz (out)	37 dBV	33 dBV	43 dBV	42 dBV
IP_3 @ 10MHz (out)	39 dBV	35 dBV	41.5 dBV	40.5 dBV
CM input range	1.4 - 3.5V	---	1.2 - 1.7V	1.15 - 1.8V

As can be seen from the table, the bipolar circuit has better linearity. The CMOS circuit has, however, better noise performance. This is partly due to the mistake when choosing the input stage of the bipolar circuit. A higher supply voltage is used for the bipolar circuit to compensate for the output voltage swing loss due to the emitter-follower. Despite that, the power consumption of the bipolar is slightly lower than for the CMOS circuit. An advantage of the CMOS circuit is that the common-mode input voltage range is much larger than for the bipolar. Two entries are missing in the CMOS measurement column. These measurements were not available in time for the publication, but it is unlikely that they differ much from the simulations.

The power consumption is not low. This can be explained by the differential topology and the requirement of low noise and distortion over a large bandwidth. With lower requirements the power consumption could be decreased. It would also be possible to achieve a power reduction without sacrificing performance by using BiCMOS. The penalty is then instead an increased cost.

The distortion was simulated and measured for a number of different amplitudes and frequencies [1,2]. In table 8.2 the distortion at $4V_{pp}$ is shown for 10MHz and 20MHz.

Table 8.2 : Simulated (S) and measured (M) THD at $4V_{pp}$ out in 1kΩ

	10MHz (S)	10MHz (M)	20MHz (S)	20MHz (M)
CMOS	0.033%	0.060%	0.033%	0.16%
Bipolar	0.015%	0.020%	0.0089%	0.013%

The requirement is about 0.03% THD at $3.75V_{pp}$ and 11.5MHz. The bipolar circuit easily fulfils this. The CMOS fulfils it in the simulations, but not in the measurements. As the CMOS circuit can not provide more than about $5V_{pp}$ output signal instead of $6V_{pp}$ before clipping, the amplitude used in the test should instead be $3.75V_{pp}*5/6=3.125V_{pp}$. At $3V_{pp}$ the measured THD is 0.035% and 0.062% at 10MHz and 20MHz, respectively. This means that the CMOS circuit is sufficiently linear when its output voltage swing is taken into account. The bipolar circuit can handle larger signals with low distortion, but it is instead noisier than the CMOS circuit.

The phase-margin is larger in the bipolar circuit, making the design more robust. With 3pF external load the margin is 70 degrees in the bipolar circuit and 30 degrees in the CMOS (in the outer loop).

8.8 Chip Photos

The chip microphotograph of the CMOS amplifier is shown in figure 8.12. The bipolar is shown in figure 8.13. The CMOS chip is 1.2 x 1.3mm and the bipolar chip is 1.2 x 1.0 mm, both including pads.

8.8 Chip Photos

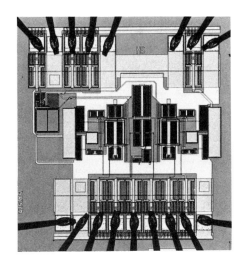

Figure 8.12: Microphotograph of the CMOS wideband IF amplifier

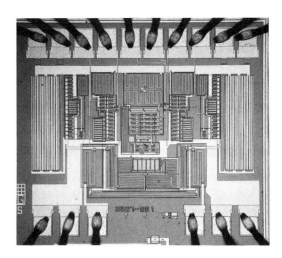

Figure 8.13: Microphotograph of the bipolar wideband IF amplifier

8.9 References

[1] H. Sjöland and S. Mattisson, 'A 100-MHz CMOS Wide-Band IF Amplifier', *IEEE Journal of Solid-State Circuits*, vol. 33, pp. 631-634, Apr. 1998

[2] H. Sjöland and S. Mattisson, 'A 160MHz Bipolar Wideband IF Amplifier', *IEEE Journal of Solid-State Circuits*, to appear Oct. 1998

[3] E. M. Cherry, 'Nested Differentiating Feedback Loops in Simple Audio Power Amplifiers', *Journal of Audio Eng. Society*, vol. 30, no. 5, pp. 295-305, May 1982

[4] E. M. Cherry, 'A New Result in Negative-Feedback Theory, and its Application to Audio Power Amplifiers', *Circuit Theory and Applications*, vol. 6, pp. 265-288, 1978

[5] R. G. H. Eschauzier and J. H. Huijsing, *Frequency Compensation Techniques for Low-Power Operational Amplifiers*, Kluwer Academic Publishers, 1995

[6] I. Hegglun, 'Square Law Rules in Audio Power', *Electronics World + Wireless World*, pp. 751-756, Sep. 1995

[7] Z. Y. Chang and W. M. C. Sansen, *Low-Noise Wide-Band Amplifiers in Bipolar and CMOS Technologies*, Kluwer Academic Publishers, 1991

Chapter 9

Inductorless RF CMOS Power Amplifiers

The purpose of an RF power amplifier is to provide gain and capability of driving an antenna at the RF frequencies of interest in the application. This chapter deals with RF power amplifiers in general, and fully integrated inductorless CMOS ones in particular.

The requirements of RF power amplifiers are discussed in the first section. CMOS RF power amplifiers using inductors are briefly treated in section 9.2. Advantages and drawbacks of inductors in CMOS technology are also discussed. Finally, in section 9.3 the design of an inductorless CMOS amplifier is described. The description is complete with measurement results and a chip photo. The amplifier is designed to investigate what performance can be achieved by a CMOS power amplifier without inductors [1].

9.1 Requirements on RF Power Amplifiers

The job of the RF power amplifier is to drive the antenna. The requirements of the amplifier often tend to be very hard to fulfil due to the high operating frequencies. In this section some of the most important requirements are discussed.

First of all, the amplifier must provide sufficient gain at the operating frequency band. This is in most cases narrow, but in some applications two or more frequency bands are used. A wideband amplifier can then be used to cover all the different bands. A drawback of narrow-band amplifiers is that they might require tuning to ensure that the correct frequencies are amplified, that is, that the gain-peak does not miss the intended band.

The amplifier must also be able to put out the amount of power required by the application. This can be substantial, resulting in the power amplifier dominating the power consumption of, for instance, a mobile telephone. High efficiency of the power amplifier is therefore often of highest importance.

In some applications a modulation scheme with non-constant envelope is used (e.g., QAM, OFDM, CDMA). By using such a modulation scheme the frequency spectrum can be used more efficiently. The drawback is that the power amplifier then must have high linearity. Nonlinearity would otherwise disturb adjacent channels due to spectral regrowth. The signal could also be distorted so that the number of bit errors increased when it was to be detected at the receiver. The demand for high linearity is particularly hard to combine with high efficiency.

It is desirable for everything in a mobile telephone to be put on a single chip using as few external components as possible. To achieve this we need an integrated RF power amplifier, preferably in CMOS and fully integrated. Building a high-performance fully-integrated RF power amplifier in CMOS is a true challenge. The MOS devices themselves are not as fast as, for instance, those in Gallium-Arsenide. In addition, coils integrated in a standard CMOS process get low quality-factors (high losses).

9.2 CMOS RF Power Amplifiers using Inductors

RF power amplifiers in CMOS using inductors have been reported [2,3]. The inductors are used to resonate with parasitic capacitances near the carrier frequency. The result is a valuable gain increase at a narrow band of frequencies. The first amplifier operates at 824-849MHz and the second at 902-928MHz.

Constant-envelope modulation is used in both [2] and [3]. The linearity is thereby not critical, so rather high efficiency can be achieved, about 40% to 50%. Integrated as well as external inductors are used in both amplifiers. The power amplifier in [3] is interesting in that it has a digital output power control. Binary-weighted NMOS-transistors connected in parallel are digitally selected using PMOS-transistors.

How to create integrated inductors in a CMOS-process is discussed in chapter 10. The coil is made as a spiral in the top metal layer or layers. Severe losses occur in the closely located conducting silicon substrate. Quality-factors of no more than about 5 can therefore be achieved in a standard process. To circumvent this, additional process steps can be added, at additional cost. Under-etching can be

9.2 CMOS RF Power Amplifiers using Inductors

used to excavate under the inductors and thereby increase the distance to the substrate [4]. Another technique is SOA (Silicon On Anything) that almost completely removes the substrate with its undesired effects [5].

As just mentioned, the main problem associated with CMOS inductors is the substrate losses which limit the quality factor (Q) at high frequencies. At low frequencies the series resistance of the metal sets the limit. At some frequency in between the Q reaches a maximum (of about 5). If the wires are made wider, the resistive losses decrease, but the substrate losses increase, so that the maximum Q occurs at a lower frequency. The selection of dimensions, such as number of turns and width of the wires, is thus very important if a reasonable Q is to be achieved at the operating frequency. The inductance is relatively straightforward to calculate [6], but the Q-factor is much more difficult. Computer programs based on, for instance, the finite-element method can be used to estimate the Q. The most accurate is to manufacture a number of inductors and then measure them. This must be repeated each time the semiconductor process is changed.

Another problem is that inductors tend to use a large chip area. This limits the inductance that can be realized on-chip. Disturbances can also be sent out and received by inductors. Inductors can for instance be used with benefit in both LNAs (low-noise amplifiers) handling extremely small signals, and power amplifiers handling the largest signals. Signals can then leak from the power amplifier to the LNA through the inductors.

A high Q is not always desired. If an inductor is used in a resonance circuit in a narrow-band amplifier, it is very important that the resonance frequency is correct for the amplifier to operate properly. The higher the Q, the sharper the resonance peak, and the more accurate the resonance frequency must be. If the Q is high, trimming might be necessary, but if it is low, the requirement is relaxed so that trimming can be avoided.

Also a low-Q inductor can be used with some benefit in an RF power amplifier. The gain can be increased by letting inductors resonate with parasitic capacitances. If an inductor is connected in parallel with a capacitor, the impedance magnitude at the resonance frequency is increased by the Q-factor. It is evident that a high Q is better, but also a Q of about 5 results in a significant gain increase. As just discussed, a low Q also makes it possible to avoid trimming.

Another advantage of inductors is that they make it possible to let the output swing around (above) the positive supply. The output voltage swing can thereby be doubled, compared to a circuit without inductors. Furthermore, an ideal inductor does not dissipate power.

9.3 An Inductorless CMOS RF Power Amplifier

An inductorless CMOS RF power amplifier was built to investigate what performance can be achieved without using inductors. A suitable topology with three stages was found. The output is single-ended to avoid an off-chip differential to single-ended transformer. As linearity is important for many modern modulation schemes, the amplifier was designed to be as linear as possible. Circuits were fabricated in a standard 0.8μm CMOS process. A 3V supply was used. The amplifier is a wideband design with an operating band from 60MHz to 300MHz (-3dB). The measured midband power gain is 30 dB with 50Ω resistive source and load impedance. The measured third order intercept point is 23 dBm and the 1dB compression point is 10dBm, both referred to the output. The power efficiency, however, is less than 15%.

9.3.1 Amplifier Topology

There are several requirements to take into account when choosing the topology of the inductorless CMOS RF power amplifier:

Since the parasitics are not well modelled at high frequencies, the amplifier must be robust. This includes stability against self-oscillation. A simple topology is therefore preferred. Since the available gain is very limited at high frequencies, feedback should be used as little as possible. The amplifier is to be a wideband design, so the frequency response is to be made flat in the operating band. Finally, the topology must be suitable for full integration; that is, external components must be avoided.

A topology trying to meet the above requirements is shown in figure 9.1.

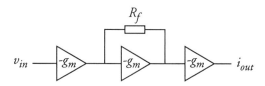

Figure 9.1: Topology of the inductorless CMOS RF power amplifier

Some feedback is used, but only over the middle stage. The feedback is employed to create a flat frequency response in the operating band. The implementation of the different amplifier stages is described in the following sections.

9.3.2 Output Stage

To achieve high linearity and large output voltage swing, a push-pull class A output stage similar to that of the CMOS wideband IF amplifier (chapter 8) is used. The stage is based on a CMOS inverter with the common-source devices enabling a large output voltage swing, see figure 9.2.

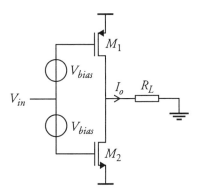

Figure 9.2: Simplified schematic of the output stage

Just as with the output stage of the CMOS IF amplifier, this stage will be linear for output currents up to 4 times the quiescent current I_q, if the devices are assumed to be matched and of ideal square-law characteristic. The stage operates in curved class A. The linearity is based on cancellation of the second order terms when two second-order polynomials are subtracted [7]. If the devices are well matched, the second order nonlinearity can be cancelled to a large extent. Real MOS devices, however, also have higher orders of nonlinearity which are not cancelled. The linearity with ideal devices can be shown by assuming V_{in} to be less than the effective gate-source voltage V_{gseff0} and the devices to be in saturation:

$$I_o = \frac{K}{2}(V_{gseff1}^2 - V_{gseff2}^2) = \frac{K}{2}((V_{gseff0} - V_{in})^2 - (V_{gseff0} + V_{in})^2)$$
$$= \frac{K}{2}(-4V_{gseff0}V_{in})$$

$$I_q = \frac{K}{2}V_{gseff0}^2 \Rightarrow I_o = -\frac{4I_q}{V_{gseff0}}V_{in} \qquad (9.1)$$

The maximum desired output current determines how large the quiescent current I_q must be. To achieve high linearity, the quiescent current is chosen as the maximum output current divided by four. According to (9.1) the stage is then ideally linear for all output currents. When the maximum current is delivered to the load, one of the transistors just turns off.

What now remains is to choose V_{gseff0}. After that everything else can be calculated. From (9.1) it is clear that the transconductance is proportional to the inverse of V_{gseff0}. The output voltage swing before the devices enter the linear region is also increased when V_{gseff0} is decreased. The reason for not selecting V_{gseff0} extremely small is that the transistor dimensions increase when V_{gseff0} is decreased:

$$I_q = \frac{1}{2}\mu C_{ox} \frac{W}{L} V_{gseff0}^2 \Rightarrow \frac{W}{L} = \frac{2I_q}{\mu C_{ox} V_{gseff0}^2} \qquad (9.2)$$

All devices are of minimum length to achieve maximum speed. The area and thereby the capacitances of the devices are then, according to (9.2), proportional to the inverse of V_{gseff0} squared. For a fixed output amplitude and frequency the current gain is then proportional to V_{gseff0}:

$$|i_{in}| \sim \frac{C}{A_v} \sim \frac{1}{V_{gseff0}} \Rightarrow A_I \sim V_{gseff0} \qquad (9.3)$$

The product of the current gain and the transconductance is thus independent of V_{gseff0}. More detailed calculations show that the product is proportional to the quiescent current I_q. Experiments could therefore be performed increasing the current and thereby reducing the requirements on the driving stage. If a larger quiescent current is used to achieve a higher current gain, the output voltage swing capability will be reduced. The linearity at lower amplitudes will, however, benefit from the larger current.

A compromise has to be made when choosing V_{gseff0}. A low value results in a large voltage gain and output voltage swing capability, but also in a low current gain and thereby efficiency. In addition, the chip area will be large. In the case of voltage drive, too low an input impedance will also be a problem. A voltage driven stage was simulated to examine how the choice of V_{gseff0} influences the distortion. The simulations were made at a low frequency using an MOS2 model. The quiescent current was 5mA in all cases and the load was 50Ω, so that the stage would operate in the curve-linear region up to 1V output amplitude. The supply voltage was 3V. The results of the simulations are presented in figure 9.3.

9.3 An Inductorless CMOS RF Power Amplifier

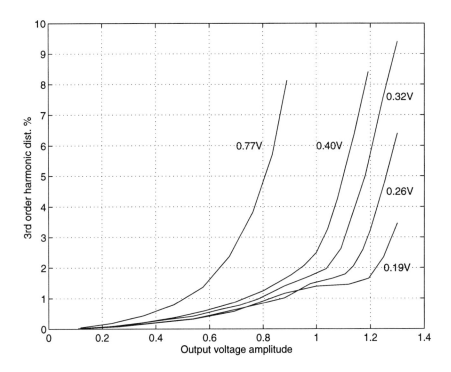

Figure 9.3: The 3rd order harmonic distortion vs. V_{gseff0} and V_{out}

The 0.77V curve corresponds to the gates being connected directly together. As expected, this results in large distortion at high amplitudes.

As linearity is important in this design, a relatively low V_{gseff0} was chosen, 0.26V. This results in a current gain equal to one at 300MHz and a voltage gain of 3.7. The efficiency would be poor at operation above 300MHz, since the current gain of the output stage is less than one at those frequencies, requiring a large bias current in the driver stage.

9.3.3 Driver Stage

The input impedance of the output stage is capacitive, so in order to move the associated pole up in frequency, the driver stage must have a low output impedance. This is achieved by a shunt feedback, figure 9.4.

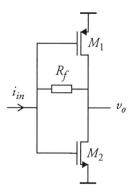

Figure 9.4: The driver stage

As the output stage has a voltage gain larger than 3, the voltage swing at the output of the driver is not critical. The gates can therefore be connected together as in an inverter.

If parasitics are not regarded, the output impedance is given by:

$$r_{out} = \frac{1}{g_{m1} + g_{m2}} = \frac{1}{g_{mtot}} = \frac{1}{\omega_{pol} c_{io}} \Rightarrow g_{mtot} = \omega_{pol} c_{io} \qquad (9.4)$$

where c_{io} is the input capacitance of the output stage. Using (9.4) the g_{mtot} required in the absence of driver stage parasitics can be calculated. When the parasitics of the driver stage and the output resistance of the previous stage, R_{oi}, are accounted for, the poles become the solutions to:

$$s^2 + s\left(\frac{g_{mtot} c_{gdtot}}{c_{gstot} c_{io}} + \frac{1}{R_f c_{io}} + \left(\frac{1}{R_f} + \frac{1}{R_{oi}}\right)\frac{1}{c_{gstot}}\right) + \frac{g_{mtot}}{R_f c_{gstot} c_{io}} = 0 \qquad (9.5)$$

The poles can be placed at appropriate complex frequencies by proper choice of g_{mtot} and R_f. In this design a Q of about 1 and a bandwidth of 280MHz were chosen. High bandwidth leads to high g_{mtot} and thereby to high current consumption. In addition, to maintain Q when increasing g_{mtot}, R_f and thereby the gain must be decreased. The g_{mtot} required to obtain a bandwidth of 280MHz is so large that the resulting quiescent current is larger than necessary to ensure class A operation of the driver stage. The quiescent current in the driver stage is thus determined by the small-signal requirements.

9.3.4 Input Stage

The input stage is to be a transconductance. An inverter is used also here, but this time without feedback. The output resistance of this stage, R_{oi}, is included in (9.5), and the output capacitance can be included by adding it to c_{gstot}. The only pole added by this stage is the one due to the input capacitance and the impedance of the signal source. This, however, occurs at a frequency much higher than the ones in (9.5), if the input stage is reasonably dimensioned and the source impedance is not excessive.

The only parameter to choose in this stage is the width of the transistors. If the transistors are wider, the gain is increased, but so is the current consumption. The output impedance also gets lower. The effect of that can be calculated using (9.5). The input capacitance also increases, so the pole due to the signal source resistance can in extreme cases become significant. The basic trade-off is, however, between gain and current consumption.

9.3.5 Simulations and Measurement Results

The entire schematic with device parameters is shown in figure 9.5. This is what was simulated, sent to fabrication and measured. The process used was a 0.8μm CMOS process with double polysilicon layers. The simulations were made using the simulator Spectre and the MOS2 transistor model. Layout parasitics and bondwire inductances were included.

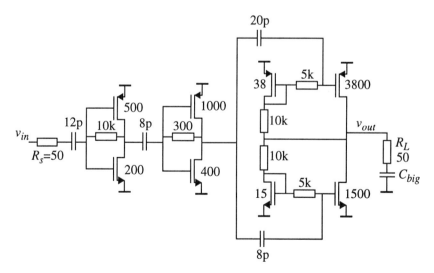

Figure 9.5: The total schematic with device parameters

As was done with the IF amplifiers, the circuits were delivered in surface mount packages. Therefore, printed circuit boards (PCB) were made for the measurements. The circuits, decoupling capacitors and contacts (SMA) for the signals were mounted on the PCBs.

The supplied voltage and current were measured all the time. The simulated quiescent current at 3V supply was 20mA. The input stage and the output stage used 5mA each, and the driver stage used 10mA. The measured current consumption, however, was 30mA at 3V. The supply voltage had to be reduced to 2.64V to get 20mA quiescent current. All the measurements are therefore made at both 3V and 2.64V.

The gain is defined as power delivered to the load divided by power available from the source. This is the same as transducer power gain [8]. Both the load and source impedance were 50Ω, as shown in the schematic. The simulated frequency response is shown in figure 9.6, where the midband gain is 29dB and the upper
-3dB frequency is 280MHz. As the amplifier is intended to be used in a system on a chip, its input impedance including encapsulation, is of low interest. Simulations were made to find the internal input impedance instead. At 300MHz it was mainly capacitive at 71-j220Ω. To measure the gain versus frequency, the PCB was connected to a network analyser. The result is shown in figure 9.7.

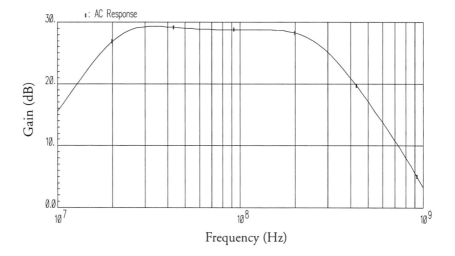

Figure 9.6: Simulated frequency response

9.3 An Inductorless CMOS RF Power Amplifier

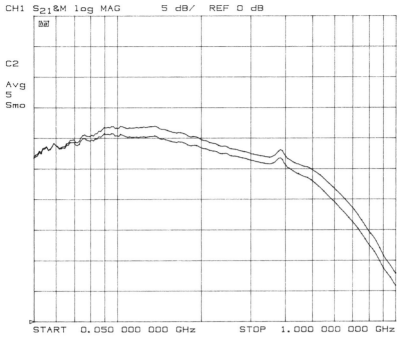

Figure 9.7: Measured frequency response. Upper curve 3V, lower 2.64V.

A two-tone test was used to simulate and measure the linearity. In the simulation the frequencies of the test tones were 200MHz and 220MHz. A transient analysis followed by an FFT was performed. The third order intercept point was then calculated to 27.5dBm referred to the output. In the measurements the frequencies were 199MHz and 201MHz. The tones were generated by two signal generators. The generators were connected to a combiner feeding the amplifier input. The amplifier output was connected to a spectrum analyser, so the amplitude of the different tones could be measured. Measurements were made at different input amplitudes, and intercept diagrams were generated, see figure 9.8. The third order intercept point was about 23dBm for both 3V and 2.64V. Compared to the simulations, this is 4.5dB less. The simulations were repeated using worst case power parameters, which resulted in about the same current consumption as in the measurements and an intercept point of 26dBm. It is therefore possible that some of the deviation between the simulations and the measurements is due to parameter variations in the fabrication. Some of the deviation can probably also be explained by imperfect transistor models.

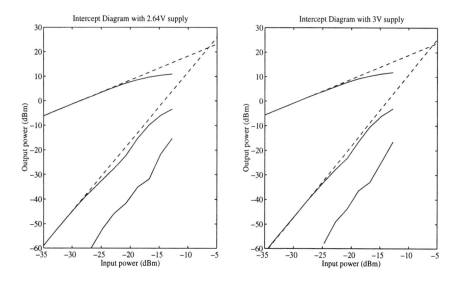

Figure 9.8: Measured intercept diagrams

As the efficiency is important, the power added efficiency (PAE) was measured. The input was connected to a signal generator, and the output to a spectrum analyser. A tone at 200MHz was used as an input signal. The output power and current consumption were measured for different input amplitudes. The PAE could then be calculated, and together with output power be plotted versus input amplitude, see figure 9.9.

The efficiency is not high, about 10% to 15% at compression, but as was mentioned before, a compromise between bandwidth, linearity and efficiency has to be made. If 300MHz operation without inductors combined with high linearity using a standard 0.8µm CMOS process is desired, the efficiency will suffer.

9.3 An Inductorless CMOS RF Power Amplifier

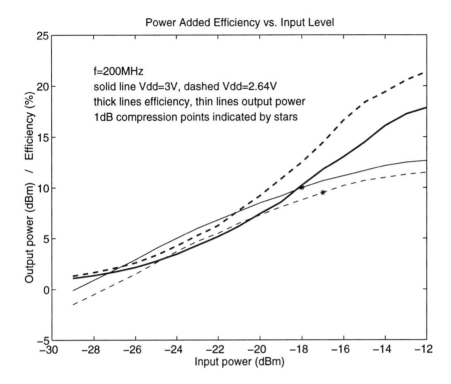

Figure 9.9: Measured power added efficiency (PAE)

9.3.6 Conclusions

A topology suitable for fully integrated CMOS RF power amplifiers is presented. Its performance was investigated for 0.8μm CMOS. An amplifier with high linearity and a measured upper -3dB frequency of 300MHz was built. The transducer power gain is about 30dB with 50Ω load and source. At 3V supply voltage, the 1dB compression point is 10dBm and the 3rd order intercept point is 23dBm, all referred to the output. The power added efficiency at compression is, however, just 10% to 15%.

When the minimum gate length decreases in future CMOS processes, the bandwidth and efficiency will increase, and mobile-phone power amplifiers operating at 2GHz might then be completely integrated in CMOS using this topology.

Since many parameters are involved, it is difficult to predict the increase in bandwidth when the minimum gate-length decreases in future CMOS processes. At large gate-lengths f_t is proportional to the inverse of the gate-length squared. Below about 0.5µm, however, f_t is instead proportional to the inverse of the gate-length itself. A reasonable estimate of the required gate-length to achieve 2GHz operation is therefore:

$$L_{estimate} = \left(\frac{0.8}{0.5}\right)^2\left(\frac{300}{2000}\right)0.5\mu m = 0.19\mu m \qquad (9.6)$$

If larger bandwidth is not needed, the reduction of the parasitic capacitances can be used to increase the efficiency instead of the bandwidth.

9.3.7 Layout and Chip Photo

The chip is 800µm x 600µm including pads. The capacitors are implemented between the two polysilicon-layers, and the transistors use short fingers contacted at both ends to achieve maximum speed. The chip photo is shown in figure 9.10.

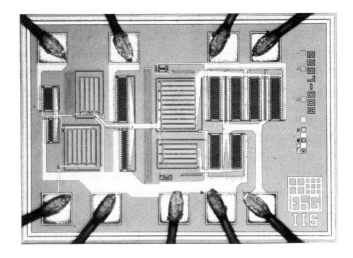

Figure 9.10: Microphotograph of the RF power amplifier

9.4 References

[1] H. Sjöland, 'An Inductorless 300MHz Wideband CMOS Power Amplifier', to appear in *Analog Integrated Circuits and Signal Processing*

[2] D. Su and W. McFarland, 'A 2.5-V, 1-W Monolithic CMOS RF Power Amplifier', *Proc. IEEE 1997 Custom Integrated Circuit Conference*, pp. 189-192, 1997, New York

[3] A. Rofougaran, G. Chang, J. J. Rael, J. Y. C. Chang, M. Rofougaran, P. J. Chang, M. Djafari, M. K. Ku, E. W. Roth, A. A. Abidi and H. Samueli, 'A Single-Chip 900-MHz Spread-Spectrum Wireless Transceiver in 1-µm CMOS Part I: Architecture and Transmitter', *IEEE Journal of Solid-State Circuits*, vol. 33, no. 4, pp. 515-534, Apr. 1998

[4] A. Rofougaran, G. Chang, J. J. Rael, J. Y. C. Chang, M. Rofougaran, P. J. Chang, M. Djafari, J. Min, E. W. Roth, A. A. Abidi and H. Samueli, 'A Single-Chip 900-MHz Spread-Spectrum Wireless Transceiver in 1-µm CMOS Part II: Receiver Design', *IEEE Journal of Solid-State Circuits*, vol. 33, no. 4, pp. 535-547, Apr. 1998

[5] P. Baltus, 'Put your power into SOA LNAs!', *Proc. AACD '98*, Copenhagen, Denmark, April 28-30, 1998

[6] H. M. Greenhouse, 'Design of Planar Rectangular Microelectronic Inductors', *IEEE Trans. on Parts, Hybrids and Packaging*, pp. 101-109, June 1974

[7] I. Hegglun, 'Square Law Rules in Audio Power', *Electronics World + Wireless World*, pp. 751-756, Sept. 1995

[8] G. Gonzales, *Microwave Transistor Amplifiers*, Prentice-Hall, 1984

Chapter 10

Layout Aspects

Before an integrated circuit can be manufactured a layout must be created. It is used to create the masks for the different fabrication steps. A proper layout is essential for a well-performing circuit. The chip area can be minimized by a carefully planned layout. Device properties such as matching, noise and high-frequency performance also rely on good layout.

Creating the layout of an analog or mixed analog/digital circuit is a highly complicated task with a large number of things that must be kept in mind. Luckily there are computer programs that can check a number of things for you. For instance, they can check if the layout violates any of the design rules of the manufacturer, or if the layout matches the schematic of the circuit. It is often easier to check the layout if it is hierarchical. The circuit is then divided into smaller parts, building blocks, which in their turn can consist of building blocks, and so on. The layout of the different building blocks can then be checked before they are combined to larger structures. A small building block can be checked fast, since the programs will run fast, and possible mistakes are easy to locate. The design rules tell, for instance, how close two objects may be located and how narrow a conductor may be regarding production tolerances and other phenomena. There are a number of design rules available from the manufacturer. If they are violated, the chances of a working circuit are reduced.

To create a good layout it is necessary to know which devices are critical for the performance of the circuit. The effort should be concentrated on those devices, making the critical parameters as good as possible. The material of this chapter can then be used. It deals with the layout of both passive and active devices, and how to achieve, for instance, high matching accuracy. The section about passive devices covers resistors, capacitors and coils. The section about active devices covers MOS and bipolar transistors, with the focus on MOS layout for small area, low noise and high speed.

10.1 Passive Devices

Passive devices are used in most analog circuits. Integrated capacitors and resistors have been used for a long time with good results. For the moment there is a large interest in making integrated coils in silicon processes, since coils have large advantages for RF circuits. The closeness to the silicon substrate, however, makes it very difficult to achieve high quality-factors (low losses).

The standard (digital) processes are optimized for high performance of the transistors. Little or no thought has been given to the passive devices, and no extra process steps are in general available for them. They can be realized using different layers of the semiconductor process that are anyway provided for the transistors. Depending on how the passive devices are realized, different performance in terms of accuracy, chip area and parasitic effects can be achieved.

10.1.1 Resistors

A layer in which resistors are to be built must have significant resistance. In principle any layer except metal can be used. (For very small resistors even metal can be used.) High resistivity is generally an advantage, as it results in low area for large resistances. Properties such as accuracy, linearity and parasitic capacitance to the substrate must also be considered.

In CMOS and advanced bipolar processes at least one layer of polysilicon, poly, is available. This is the layer most suited for high-performance resistors. It has the highest linearity, low parasitic capacitance and moderate temperature drift. This layer is used for the resistors of the amplifiers in this book. The absolute accuracy is poor, however, often about 25%. Another drawback is the low resistivity of about 20Ω per square.

Another option is to use diffusion resistors. In a CMOS process the source/drain diffusion is then used. The resistivity and temperature drift are about the same as for poly. The resistors will, however, be more nonlinear and have a larger parasitic capacitance to the substrate, which makes poly a better choice for MOS. Bipolar technology provides many more alternatives for diffusion resistors [1]. Base diffusion, emitter diffusion and pinched base diffusion resistors can be implemented. The same problems apply to them, but the pinched base diffusion has the advantage of a resistivity of about $10k\Omega$ per square.

10.1 Passive Devices

For high resistivity the well of a CMOS process or the epitaxial layer of a bipolar process can be used. The resistance is about 10kΩ per square. The different resistor realisations with approximate figures of merit are described in [1]. The drawbacks of well-resistors are large nonlinearity and parasitic capacitances.

In CMOS it is also possible to use an MOS transistor in the linear region as a resistor. The resistance between drain and source can be controlled by the gate voltage. Very area-efficient resistors can be built in this way. The drawback is large nonlinearity.

When the layer has been chosen it is time to draw the layout. An area-efficient resistor layout is shown in figure 10.1. The resistor is shaped like a snake.

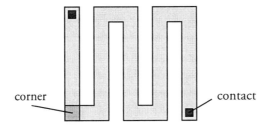

Figure 10.1: Example of resistor layout

To calculate the resistance, the effective length and width of the resistor is used:

$$R = \frac{L_{eff}}{W} R_{sq} \qquad (10.1)$$

where R_{sq} is the resistance per square. The effective length L_{eff} is tricky to calculate in the corners, but according to [2] a corner has an effective length equal to 0.56 times W.

Often the relative accuracy (matching) of two or more resistors is critical. Special measures then have to be taken when drawing the layout. The matched resistors are made from a number of identical resistor elements, which are placed with the same orientation. Dummy resistor elements can be used in order to create almost identical surroundings of the resistor elements in use. Metal interconnects are then used to create resistors with a common centre of gravity (common centroid). The common centre of gravity is used for all matching, not just for resistors. The idea is that a gradient in, for instance, layer thickness will affect all devices with the same centre of gravity equally. An example of two matched resistors is shown in figure 10.2.

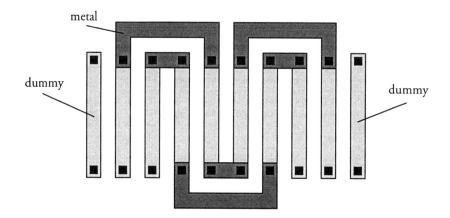

Figure 10.2: Layout of two matched resistors

If good matching is required, the resistors must be wide enough, as the influence of width fluctuations then is decreased. A larger area also has an averaging effect on fluctuations of layer resistivity. There is thus a conflict between small area and good matching.

The parasitics involved are mainly capacitances to the substrate. They result in an upper frequency limit above which the resistor can not be used. They can also pick up undesired signals from the substrate. The parasitic capacitance is distributed along the resistor, but in most cases sufficient accuracy is achieved by a model with half the capacitance at each end, see figure 10.3.

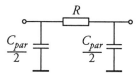

Figure 10.3: Simple model of resistor with parasitic capacitance

10.1.2 Capacitors

Just as for resistors, there are a number of different ways to implement capacitors. A capacitor is usually implemented using two conducting layers separated by silicon-oxide. The thinner the oxide, the larger the amount of capacitance per area that can be implemented. The thinnest oxide in a CMOS process is the gate-oxide. Area-efficient capacitors can therefore be implemented using MOS transis-

10.1 Passive Devices

tors. They are biased in strong inversion. The source and drain are used as one terminal and the gate as the other. The drawback is that the capacitance is highly voltage dependent (nonlinear). In some bipolar processes, MOS-capacitors are available. The doping levels are then chosen for high linearity, resulting in excellent capacitors. In some CMOS processes a second polysilicon-layer is available. Excellent capacitors can then be implemented between the two poly-layers, which are separated by an oxide with a thickness comparable to that the gate-oxide.

It is also possible to use the capacitances between the interconnect layers. A problem is that these normally are regarded as parasitics which hence are to be minimized. The oxide separating the interconnect layers is therefore rather thick. To achieve a reasonable capacitance per area, several interconnect layers can be used in a sandwich structure. Modern processes tend to have several metal layers, which is a result of the demands from primarily digital circuits.

In both CMOS and bipolar technology it is possible to create diodes, which can be used as capacitors. The capacitance becomes strongly voltage dependent. This is a drawback in amplifiers as the linearity is affected, but in applications such as voltage controlled oscillators (VCOs) this is a desired effect.

Just as for resistors, the matching of capacitors can be critical. An example of two matched capacitors is shown in figure 10.4.

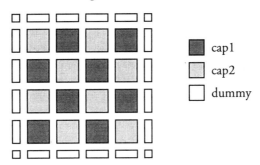

Figure 10.4: Layout of two matched capacitors

Just as for the resistors, the devices are split into identical pieces, which then are placed so that the two devices get a common centre of mass. Also here the structure is surrounded by dummies. They can be skipped if the requirements are modest. The figure does not show the connections of the capacitors. Care must be taken when creating them, as their capacitance otherwise can spoil the matching. Just as for resistors, large devices result in better matching, thanks to statistical averaging.

There are mainly two types of parasitics that must be considered. The first is parasitic capacitance from the bottom plate of the capacitor to the layer below, most often the substrate. The bottom plate of the capacitor must be connected to a low-impedance node in order to minimize the effect of this capacitance. The second effect is series resistance. In for instance a poly-poly capacitor, the resistivity of the poly-layers might cause a significant series resistance. The layout must be created with this in mind; that is, the maximum distance to a contact to metal must be short enough. An MOS gate-capacitor must in the same way also be short enough. How large a resistance can be allowed depends on the frequency of operation. The higher the frequency, the lower must the resistance be not to dominate the impedance. A simple capacitor model with parasitics is shown in figure 10.5.

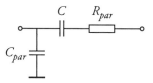

Figure 10.5: Simple model of capacitor with parasitics

10.1.3 Inductors

Integrated inductors, particularly in CMOS, are an area of large current interest [3-6]. The reason is the benefits of inductors for RF circuits, combined with the efforts for building RF circuits in CMOS.

It is hard to create inductors with high quality (low losses) in standard processes. CMOS tends to be worse than bipolar in this respect, depending on the heavier doping of a typical CMOS substrate. The closeness to the conducting substrate causes losses. Even the top metal layer is not more than a few microns away. Other losses occur due to the resistance of the coil windings, which at high frequencies is increased due to the skin effect.

There are not so many layers to choose from when creating an inductor. To achieve reasonable performance the top metal layer/layers must be used. The coil is then drawn in the shape of a spiral, see figure 10.6.

10.1 Passive Devices

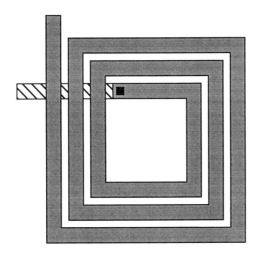

Figure 10.6: Layout of an integrated inductor

The spiral must not be square as in figure 10.6. If the layout rules permit, a circular shape is best. Octagonal inductors can be drawn in many processes, and their performance is not far from that of the circular ones. The coil windings should not continue to the centre of the coil. A centre-hole should instead be left out as in figure 10.6. Windings in the middle do not contribute much to the inductance, but instead to the series resistance.

A number of layout parameters are critical for the performance of the inductor. One such parameter is the number of turns; others are the conductor width and spacing. How to select these parameters depends on the frequency at which the coil is to be used. A high frequency results in a smaller area than a low one. Even if the coil is well designed, it is hard to achieve a quality factor (Q) of more than five to ten in a standard CMOS process.

It is hard to predict the performance of an integrated inductor. The inductance can be calculated using the method in [7], but the Q is more difficult to find. More or less crude estimates can be made using hand-calculations. To get a more accurate estimate, three-dimensional computer simulation can be used.

To increase the Q and reduce the signal coupling through the substrate, experiments have been made with shielding [4]. The best results were achieved with patterned shields of polysilicon. The pattern was designed to reduce the currents induced in the shield.

It also is possible to increase the Q by non-standard process steps. For instance, chemical etching under the inductors can be used to increase the distance to the substrate. Another possibility is to etch the entire substrate away and replace it with, for instance, glass, as in an SOA-process (Silicon On Anything). The additional process steps, however, increase the cost.

If a sufficient Q can not be achieved using integrated inductors, bondwires can be used instead. The inductance spread is, however, much larger than for integrated inductors, which in fact are among the most accurately reproducible devices in an IC process. A lower spread can be achieved using on-chip bondwires, if such are available.

10.2 Active Devices

Two different active devices are used in this book, the MOS-transistor and the bipolar transistor. The layout of the transistors is more complicated than of the passive devices, as several different layers (fabrication steps) are involved. The focus of this section will be on layout of high-performance MOS-transistors. A brief discussion on bipolar transistors is also included.

10.2.1 MOS Transistors

Proper layout is important in order to achieve high performance transistors. The most commonly encountered performance requirements are high speed, low noise and small area. Sometimes high current-handling capacity or high matching accuracy is also needed. It is not possible to achieve maximum performance in all aspects at the same time. This is, however, not necessary since one transistor typically needs maximum performance in just one or two aspects. The solution is therefore, depending on the requirements, to make the layout different for different devices. In this section MOS-transistor layout for different requirements is described.

10.2.1.1 The Layers and Their Physical Correspondence

As mentioned, several fabrication steps are needed to produce a transistor, and thereby several layers are involved when creating the layout. To get an understanding of the layers involved and their physical correspondence, a layout of a simple NMOS-transistor together with the cross-section of the corresponding physical device is shown in figure 10.7.

10.2 Active Devices

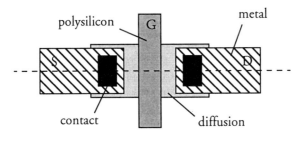

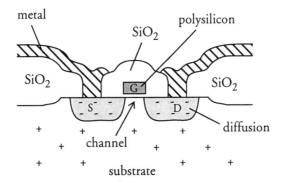

Figure 10.7: Simple NMOS transistor layout and the corresponding cross-section of the physical device

The example is for a process using a p-doped substrate. If a PMOS-device is to be made in such a process, an n-doped well must be created in which the transistor can be made. Furthermore, the diffusion is to be p-doped instead of n-doped, and this is often accomplished by an additional layer in the layout. Apart from the extra layers, however, the layout of the devices is the same.

As can be seen in figure 10.7, the active part of the transistor, the channel, is located at the intersection of the polysilicon and the diffusion layer. This intersection thereby defines the width and length of the device. The purpose of the rest of the structure is to provide the active part with signals. It also introduces parasitic capacitances, however, and minimizing this by clever layout not only reduces the area but also possibly increases the speed of the device.

10.2.1.2 Layout for Low Noise

The noise of a transistor can be divided into two categories. First there is the noise generated in the active part of the transistor, the channel. This noise is unaffected by the layout, as long as the width and length of the transistor are not changed. The second category is the noise generated by layout-dependent resistances in series with the transistor terminals. A low-noise transistor should have a layout that makes the noise of the second category considerably smaller than that of the first.

The most important layout-dependent resistance concerning noise is the one in series with the gate. Modern CMOS processes use polysilicon gates. This material has a rather high resistivity compared with metal. In combination with the narrow gates (short channels) of high-performance devices, this calls for attention when the layout is drawn. To get a low resistance, the gate must be connected as densely as possible to metal.

Another source of noise is the bulk resistance. The noise of the bulk affects the drain current through g_{mb}. Since g_{mb} is much smaller than g_m, the resistance of the bulk is allowed to be considerably larger than the resistance of the gate. To get a low enough bulk resistance, the transistor should be surrounded as closely as possible by a metal ring with bulk contacts. This also reduces the disturbances from other parts of the circuit, which can be regarded as noise.

To summarize: To achieve low noise, the gate should be connected as densely as possible to metal, and the transistor should be surrounded as closely as possible by a metal ring with bulk contacts.

10.2.1.3 Layout for High Speed

Just as with noise, the speed limiting phenomena can be divided into two categories. One is the parasitics associated with the active part of the transistor, which is unaffected by the layout. The other is the layout-dependent parasitic capacitances between the transistor terminals. Also, the parasitic resistance in series with the gate can affect the speed, and therefore belongs to this category.

Since the gate resistance can affect the speed, it should be minimized. Just as for low noise, this is done by connecting the gate as densely as possible to metal. To accurately predict the influence of gate-resistance, the gate should be modelled as a distributed RC-network.

To minimize the parasitic capacitances from drain and source to bulk, the area of the drain and source diffusions should be minimized, which also is advantageous for the total area. High-speed transistors are often used with bulk and source

10.2 Active Devices

connected to ground. The capacitance associated with the source diffusion does not then affect the performance. The effort can then be put entirely on minimizing the area of the drain diffusion. The result is non-symmetric transistors with different areas of source and drain.

10.2.1.4 Layout Techniques for Wide Transistors

There are different alternatives for the layout of high-performance CMOS transistors with large W/L-ratios. Such transistors are very common in analog circuits, and a simple, yet effective, way of implementing them is to use the finger layout, figure 10.8.

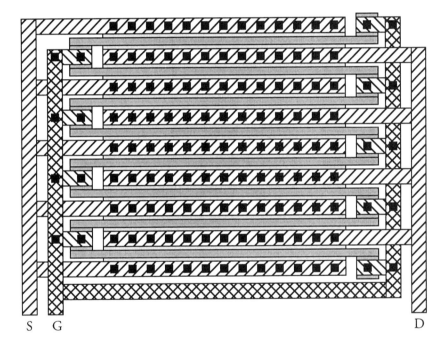

Figure 10.8: Example of finger layout

The wide transistor is made up of a number of smaller devices that are connected in parallel. In this way, even a very wide transistor can get an area-efficient square shape. Another advantage is that the drain and source diffusions have channels at two sides, which approximately halves the diffusion area needed. Furthermore, the structure allows the gate to be contacted densely. In the figure the gates are contacted to metal at both ends for lowest noise.

To use this layout technique, the number of parallel devices (fingers) must be chosen. A large number of fingers results in low gate resistance, and thereby low noise and possibly high speed, but the area will not be minimal. As usual, a compromise must be found.

Thanks to the simple structure, a transistor with desired width and length can quickly be drawn. It is also relatively easy to perform calculations of, for instance, the gate resistance, as the structure is just a parallel connection of simple devices. This layout technique has been used for all the wide MOS transistors of the amplifiers in this book.

Although the finger layout has a lot of advantages, there are competitive alternatives. One is the waffle-iron layout, figure 10.9.

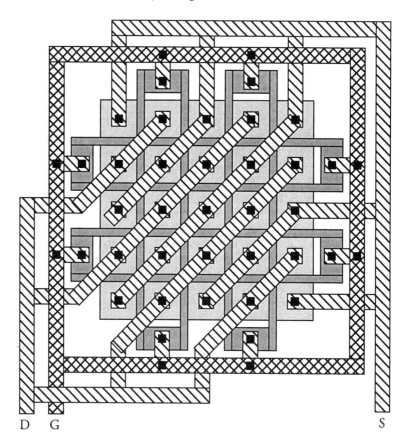

Figure 10.9: Example of waffle-iron layout

10.2 Active Devices

The idea is to surround the drain and source diffusions with a channel on all four sides, instead of just on two as in the finger layout. A comparison of the finger versus the waffle-iron layout is found in [8]. Which layout technique is most area-efficient depends on the design rules. The large number of parallel paths through the mesh of gates results in a lower gate resistance for the waffle-iron technique. This makes it interesting for low-noise devices.

10.2.1.5 High Current Layout

In, for instance, the output devices of the audio power amplifiers in chapter 7, the drain current can be several hundreds of milliamperes. This must be regarded when the layout is drawn, otherwise the large current will cause failure of the circuit. A typical metal-layer in a CMOS process can not stand more than about one milliampere per micrometer conductor width. This means that a metal connector must be about 0.25mm wide to carry 250mA. The area consumed by connectors in high current circuits is thus very large, and thereby worth trying to minimize.

To minimize the conductor width, as many of the available metal-layers as possible should be used in parallel. If there are two metal layers available, the conductor width can in this way be reduced to about one half, and if three layers are available to one third, and so on. An example of how this principle can be employed in a high current transistor, implemented in a process with two metal layers, is shown in figure 10.10. A similar layout was used to create area-effective output devices in the 1.5V audio power amplifier in chapter 7.

10.2.2 Bipolar Transistors

The chip manufacturer usually provides a set of different bipolar transistors complete with layout and simulation models. The reason is that bipolar transistors are hard to model accurately from a user defined layout. In this way the circuit designer does not have to know anything about the details of bipolar transistor layout. One has only to select suitable transistors and use the layout supplied by the manufacturer.

In high current applications it is often necessary to connect a large number of devices in parallel, not to exceed the maximum allowed current density.

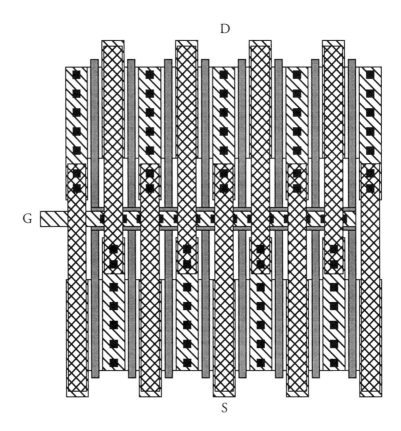

Figure 10.10: High current MOS transistor in a process with two metal layers

10.2.3 Matching of Transistors

Matching active devices is based on the same principle as with passive devices. If two or more transistors are to be matched, they should have the same centre of gravity. This is called the common-centroid technique [9]. The principle is illustrated for two capacitors in figure 10.4.

If the finger layout is used for a pair of MOS-transistors, every other drain-finger can belong to one or the other transistor. The result is very area-efficient set of matched transistors. This technique requires the transistors to have a common source node, which they have in most cases where matched transistors are needed, such as in differential pairs and current mirrors.

10.3 References

[1] P. R. Gray and R. G. Meyer, *Analysis and Design of Analog Integrated Circuits, third edition*, Wiley, 1993

[2] T. H. Lee, *The Design of CMOS Radio-Frequency Integrated Circuits*, Cambridge University Press, 1998

[3] K. T. Christensen and A. Jørgensen, 'Easy Simulation and Design of On-Chip Inductors in Standard CMOS Processes', *Proc. IEEE International Symposium on Circuits and Systems*, pp. 360-364, vol. IV, 1998

[4] C. P. Yue and S. S. Wong, 'On-Chip Spiral inductors with Patterned Ground Shields for Si-based RF IC's', *IEEE Journal of Solid-State Circuits*, vol. 33, no. 5, pp. 743-752, May 1998

[5] C. P. Yue, C. Ryu, J. Lau, T. H. Lee and S. S. Wong, 'A Physical Model for Planar Spiral Inductors on Silicon', *Proc. IEEE International Electron Devices Meeting*, pp. 155-158, 1996

[6] J. R. Long and M. A. Copeland, 'The Modeling, Characterization, and Design of Monolithic Inductors for Silicon RF IC's', *IEEE Journal of Solid-State Circuits*, vol. 32, no. 3, pp. 357-369, March 1997

[7] H. M. Greenhouse, 'Design of Planar Rectangular Microelectronic Inductors', *IEEE Trans. on Parts, Hybrids and Packaging*, vol. PHP-10, no. 2, pp. 101-109, June 1974

[8] S. R. Vemuru, 'Layout Comparison of MOSFETs with Large W/L Ratios', *Electronic Letters*, vol. 28, no. 25, pp. 2327-2329, Dec. 1992

[9] M. Ismail and T. Fiez (Editors), *Analog VLSI – Signal and Information Processing*, McGraw-Hill, 1994

Index

A
absolute accuracy 10
accuracy 26
active feedback 98
amplifier applications 1
asymptotic-gain model 35
audio power amplifiers 6, 87

B
bandwidth 11
base station 3
BiCMOS 10, 90
bipolar transistors 10, 21, 157
Blomley topology 92
Bode diagram 37

C
cancellation 31
capacitor layout 148
CDMA 2, 130
chip area 11
class AB bias control 57
class of operation 52
clipping 12, 71, 72
CMOS audio power amp 6, 91
CMOS inverter 31, 57, 116, 133
common-centroid 147, 158
common-mode feedback 115, 121
compression point 13, 132
cost 9
cross-over distortion 56, 94
current density 157
current splitter 92
curved class A 53, 56

D
degeneration 119
design rules 145
device models 14
differential topology 32, 114
diffusion resistors 146
dominant pole 39
dynamic nonlinearity 68

E
efficiency 12
equivalent noise source 11

F
feedback bias controls 57
feedback boosting 27, 42
feed-forward 27
feed-forward bias control 57
FFT 65
finger layout 155

G
gain margin 37
gain-bandwidth product 46
gate-oxide 148
Gaussian 70
global feedback 41

H
harmonic content 74, 75
high-current layout 157
high-speed layout 154
hot electron noise 19

I
ideal amplifier 1, 11
IF amplifier 3
inductor layout 150
inductorless 132
inductors 130
input impedance 11
instability 11
integrated analog electronics 9
integrated circuits 9
intercept diagram 13, 139
intercept point 13, 139
intermodulation distortion 4, 65
IS-54 2

L
layout 145
level-shift 96, 103
linear region 15
linearisation techniques 25
linearised bandwidth 29
linearity requirements 4
LNA 131
load capacitance 6
load stabilizing network 97
local feedback 41
loop gain 26, 36
low voltage 56
low-noise layout 154

M
matching 10, 45, 145

matching of capacitors 149
matching of resistors 147
matching of transistors 158
maximum efficiency 55
Miller compensation 39
minimum phase 37
mobility 19
model amplifier 30
modulation 51
MOS current mirror 29
MOS transistor 14, 152
move-pole 39

N
negative feedback 7, 26
negative resistance 43
nested Miller compensation 40, 114
noise 11, 18, 23, 118, 145
noise figure 11
non-constant envelope 2, 130
non-ideal characteristics 11
nonlinearity 11
Nyquist diagram 36

O
OFDM 2, 130
output impedance 11
output voltage swing 57

P
package parasitic impedances 10
PAE 140
parallel compensation 40
passive device layout 146
passive devices 9
PCB 10
phantom zero 39, 97
phase margin 37, 126
phase-compensation 27, 39, 97
point feedback 41
pole split 39
polysilicon resistors 146
power consumption 11
power efficiency 52

Index

predistortion 29
probability density function 69
PSRR 14
pulse-width modulation 55, 90
push-pull operation 57, 60

Q

quality-factor 131, 150
quiescent current 25, 51

R

relative accuracy 10
resistor layout 146
reversed nested Miller 99
RF power amplifier 5, 129

S

saturated region 15
self-oscillations 27, 36
short channel effects 19
shunt feedback 135
signal current 51
signal splitter 92
simulation 65
slew-rate 12, 56, 77
SNR 11
SOA 131, 152
spectral density 71
spectral widening 5
spectrum analysers 66
standard digital CMOS 10
static nonlinearity 68
strong inversion 15

T

tail currents 52
TETRA 2
THD 2, 4, 12, 67
THD meters 66
thermal voltage 22
threshold voltage 14
TIM 12
transconductance 17, 22
transistor layout 152
transit frequency 17, 22

translinear circuits 10, 29
tuned amplifier 55
two-tone measurement 12, 139

U

ultimate lower supply voltage limit 57
under-etching 130, 152

V

velocity saturation 19
Volterra analysis 12

W

waffle-iron layout 156
weak inversion 15
wide transistor layout 155
wideband IF amplifiers 2, 3, 113
wideband signal 65